AUTODESK® ROADWAY DESIGN
FOR INFRAWORKS 360™

ESSENTIALS

AUTODESK® ROADWAY DESIGN FOR INFRAWORKS 360™

ESSENTIALS

Eric Chappell

Senior Acquisitions Editor: Willem Knibbe
Development Editor: Sara Barry
Technical Editor: Tommie Richardson
Production Editor: Eric Charbonneau
Copy Editor: Kim Wimpsett
Editorial Manager: Pete Gaughan
Vice President and Executive Group Publisher: Richard Swadley
Associate Publisher: Chris Webb
Book Designer: Happenstance Type-O-Rama
Compositor: Cody Gates, Happenstance Type-O-Rama
Proofreader: Rebecca Rider
Indexer: Ted Laux
Project Coordinator, Cover: Todd Klemme
Cover Designer: Wiley
Cover Image: Courtesy of Eric Chappell

Dear Reader,

Thank you for choosing *Autodesk Roadway Design for Infraworks 360 Essentials*. This book is part of a family of premium-quality Sybex books, all of which are written by outstanding authors who combine practical experience with a gift for teaching.

Sybex was founded in 1976. More than 30 years later, we're still committed to producing consistently exceptional books. With each of our titles, we're working hard to set a new standard for the industry. From the paper we print on to the authors we work with, our goal is to bring you the best books available.

I hope you see all that reflected in these pages. I'd be very interested to hear your comments and get your feedback on how we're doing. Feel free to let me know what you think about this or any other Sybex book by sending me an email at contactus@sybex.com. If you think you've found a technical error in this book, please visit http://sybex.custhelp.com. Customer feedback is critical to our efforts at Sybex.

Best regards,

Chris Webb
Associate Publisher
Sybex, an Imprint of Wiley

To my favorite company: Autodesk

Acknowledgments

First and foremost I have to thank Willem Knibbe for making this book even happen in the first place. Next I would like to thank Pete Gaughan for stepping in, taking the reins, and being helpful, decisive, positive, and just wonderful to work with. Next I would like to thank Tom Richardson for being the most thorough and utterly brutal technical editor the world has ever known. I may have said some bad words in my head about you while I was working through your comments, but in the end I couldn't be more grateful for your efforts. In addition, you over-came a lightning-fast software development process that had everyone guessing as to which version we were supposed to be using at any given time. And last but certainly not least, thanks to Sara Barry and Eric Charbonneau for your tireless efforts correcting my writing and making it so much better.

It has been a crazy winter and spring, not just for me but also for my family. By the time you read this, I will have updated one full-size book and written another full-size book and at least two mini-books like this one—all from scratch. On top of that, I started a new job with the best company in the world—a job that is fun, challenging, and also quite consuming of my time and energy. My wife and four children have endured all of it right along with me: coping with many late nights, dealing with Dad being low on sleep, dealing with Dad's distant stare while calcu-lating how the next few deadlines will play out, and enduring all the other not-so-glamorous parts of having an author in the family. I hope they're proud of me, and I hope they enjoy and benefit from my authoring efforts because when it all comes down to it, I do it all for them.

About the Author

Eric Chappell has been working, teaching, writing, and consulting in the world of civil engineering software for more than 20 years, and he is a recognized expert in the world of Autodesk® InfraWorks software. Eric joined the Autodesk family in September 2013 as a Premium Services Specialist. In the 12 years prior to that, he wrote training materials and performed training for end users, trainers, and Autodesk employees around the globe. For several years, he has worked with Autodesk in authoring and developing two Autodesk certification exams. He also served as design systems manager for Timmons Group, a civil engineering and surveying firm based in Richmond, Virginia, where he managed software, standards, and training for more than 200 users. Eric is also a highly rated instructor at Autodesk University, where he has taught for the past 10 years.

Prior to writing and consulting, Eric spent nearly 10 years in the civil engineering and surveying fields while working for the H. F. Lenz Company in Johnstown, Pennsylvania. During his time at H. F. Lenz, he gained considerable practical experience as a survey crewman, designer, engineer, and CAD supervisor. Eric also has a B.S. degree in civil engineering technology from the University of Pittsburgh at Johnstown and is certified in Pennsylvania as an EIT.

Eric is originally from southwestern Pennsylvania, but he has lived in the Richmond, Virginia, area for the past 13 years with his wife and four children. He enjoys being outdoors and spending time with his family. He can sometimes be seen playing drums for the band Sons of Zebedee, which plays at a variety of events in the Central Virginia area.

If you would like to contact the author regarding comments or suggestions, please email InfraWorksEssentials@gmail.com. You are also welcome to visit Eric's blog at http://ericchappell.blogspot.com.

Contents at a Glance

CONTENTS

INTRODUCTION

This year I have had the wonderful privilege to write about what I think is the most exciting program to come out of Autodesk in the last decade—InfraWorks. After 20 years of plugging away with AutoCAD-based Civil 3D and then Land Desktop before that, InfraWorks is a fantastic contrast: fun, stunning, and beautiful like a gaming environment. And on top of all that it is useful, practical, and powerful as a preliminary design and planning tool. As if that isn't enough, I get to keep telling the story, the three-part sequel comprised of the add-on modules Roadway Design, Bridge Design, and Drainage Design.

With other similar books that I've written, I've tried to stick to certain criteria that I think are necessary to make a good learning resource and companion. I've tweaked these for Roadway Design for InfraWorks 360 as follows:

▶ It should be basic enough to enable anyone to learn Roadway Design for InfraWorks 360.

▶ It should be in-depth enough to enable a person to be productive using Roadway Design for InfraWorks 360 for basic tasks.

▶ It should foster understanding by associating the things you do in Roadway Design for InfraWorks 360 with familiar things that you see every day.

▶ The examples and exercises should be based on the real world.

▶ The book should not simply demonstrate random software features but should teach the process of project completion using Roadway Design for InfraWorks 360.

While writing each chapter of this book, I have tried to meet these criteria. I hope before opening this book, you have gone through *Autodesk InfraWorks and InfraWorks 360 Essentials* and have become familiar with the concept of InfraWorks, its user interface, and its environment. Just as Roadway Design for InfraWorks 360 is built upon InfraWorks 360, this book is built upon its counterpart. In addition, if you've experienced *Autodesk InfraWorks and InfraWorks 360 Essentials*, then you're also familiar with the Bimsville Bypass project. In this book, you'll move the conceptual design of the Bimsville Bypass highway into a more detailed state using the powerful engineering-based tools of Roadway Design for InfraWorks 360. You'll also do the same with the road passing through the industrial park that was also created as part of *Autodesk InfraWorks and InfraWorks 360 Essentials*.

THE FULL INFRAWORKS ESSENTIALS LIBRARY

This ebook is part of a complete library covering InfraWorks. All are available at sybex.com and through a variety of online resellers.

Autodesk InfraWorks and InfraWorks 360 Essentials   **Available in print and as an ebook**

Autodesk Roadway Design for InfraWorks 360 Essentials   **Available as an ebook**

Autodesk Drainage Design for InfraWorks 360 Essentials   **Available as an ebook**

Autodesk Bridge Design for InfraWorks 360 Essentials   **Available as an ebook**

Who Should Read This Book

This book should be read by anyone who already understands the core InfraWorks features and is ready to move on to more advanced, detailed, engineering-based road design. It is designed to be used in a formal classroom environment, in your home office learning on your own, and everywhere in between. It is appropriate for ages ranging from high school to retirement, and although it is intended for those who have no experience or skill with Roadway Design for InfraWorks 360, it can also serve as a great resource for refreshing one's knowledge base or for filling in any gaps.

Here are some specific examples of individuals who would benefit from reading this book:

► Anyone involved in the planning or preliminary design stages of infrastructure or land development projects

► High-school students following a design-related educational track

► College students learning to be designers, engineers, or GIS professionals

▶ Employees who have recently joined a company that utilizes InfraWorks 360

▶ Employees who work for companies that have recently implemented InfraWorks 360

▶ Experienced InfraWorks 360 users who are self-taught and who want to fill in gaps in their knowledge base

▶ Civil 3D users

What You Will Learn

The goal of this book is to make you productive with Roadway Design for InfraWorks 360 in a relatively short time. It begins by giving you a detailed tour of the user interface and the environment you will be working in. It then covers nearly all of the Roadway Design for InfraWorks features using relatively simple but realistic examples. It takes you through the progression of a project (the Bimsville Bypass) so that you can learn the logic of how Roadway Design for InfraWorks 360 is applied to make projects more successful.

After completing this book, you will be able to use Roadway Design for InfraWorks 360 to work on the more detailed, engineering-based aspects of road design. You will be able to work on a team that also uses Roadway Design for InfraWorks 360, and you will be able to create, modify, analyze, optimize, and share roadway design models with them.

FREE AUTODESK SOFTWARE FOR STUDENTS AND EDUCATORS

The Autodesk Education Community is an online resource with more than 5 million members that enables educators and students to download—for free (see the website for terms and conditions)—the same software used by professionals worldwide. You can also access additional tools and materials to help you design, visualize, and simulate ideas. Connect with other learners to stay current with the latest industry trends and get the most out of your designs. Get started today at www.autodesk.com/joinedu.

What You Need

The following sections highlight the key things you will need to successfully complete the entire book.

Software

To work through the exercises in this book, you will need to have InfraWorks 360 installed on your computer, and Roadway Design for InfraWorks 360 must be enabled. You should refer to the Autodesk website (www.autodesk.com) and make sure your computer's specifications are equal or better than the recommended system requirements for the current release of these products.

An InfraWorks 360 Account

For Roadway Design for InfraWorks 360 to be enabled, you must have InfraWorks 360 installed. Autodesk InfraWorks (not 360) does not have the ability to accept Roadway Design for InfraWorks as an add-on program. You must also be entitled to Roadway Design for InfraWorks 360 either as a trial version or through your Autodesk account. If you work for a company or organization, check with a software manager within your organization about getting set up with the proper software and entitlements. Or, if you do not have such a person in your organization, contact your Autodesk reseller. You can also visit www.autodesk .com/infraworks for more information. Also, one of the exercises in this book requires cloud credits, so you should consult the appropriate person in your company or organization, or work with your Autodesk representative to find out whether you have access to cloud credits.

Exercise Files

To complete the exercises, you will need to download the necessary files from www.sybex.com/go/roadwayessentials. Here you will find a list of ZIP files, one for each chapter, which you should unzip to the local C: drive of your computer. This will create a folder named InfraWorks Roadway Essentials with the chapter folder inside it. As you unzip additional chapter files, simply merge the new InfraWorks Roadway Essentials folder into the old one.

Configure Your Units

When you begin using InfraWorks, you will need to configure it for either imperial or metric units. Here are the steps:

1. Launch InfraWorks 360.

2. On the start page, click Application Options.

3. In the Application Options dialog, click Unit Configuration.

4. For Default Units, select either imperial or Metric based on the units you prefer to work in.

Know Which Units to Use

As you work through the book you will see dimensions provided twice: once in imperial units and then again as metric units in parentheses. Please note that the metric value is not a direct conversion of the imperial value but is usually a similar value that has been rounded to be more practical. The following is an example step from an exercise:

9. Click the height gizmo and enter a value of **150 (45)** in the tooltip. Press Enter. Press Esc to clear the selection of the building.

In this example, you would enter 150 if you are working in imperial units and 45 if you are working in metric units. Note that 45 meters is not the exact metric equivalent of 150 feet but a practical approximation.

Check for Updates

Finally, be sure to check the book's website, www.sybex.com/go/roadwayessentials, for any updates to this book should the need arise. You can also send questions and comments directly by email at InfraWorksEssentials@gmail.com.

> **THE ESSENTIALS SERIES**
>
> The Essentials series from Sybex provides outstanding instruction for readers who are just beginning to develop their professional skills. Every Essentials book includes skill-based instruction with chapters organized around projects rather than abstract concepts or subjects, plus digital files (via download) so you can work through the project tutorials.

What Is Covered in This Book

Autodesk Roadway Design for InfraWorks 360 Essentials is organized to provide you with the knowledge needed to master the basics of the Roadway Design module.

Chapter 1: Getting Started   Familiarizes you with the concept of Roadway Design for InfraWorks 360 and gives you a detailed account of all the tools available through its user interface

Chapter 2: Designing Roads   Demonstrates how to create and modify roads addressing a number of road design aspects and utilizing a number of tools

Chapter 3: Using Advanced Functions   Demonstrates advanced functions that address optimization, analysis, and output to Civil 3D

How to Contact the Author

I welcome feedback from you about this book or about books you'd like to see from me in the future. You can reach me by writing to InfraWorksEssentials@gmail.com. For more information about my work, please visit my blog at http://erichappell.blogspot.com.

Sybex strives to keep you supplied with the latest tools and information you need for your work. Please check its website at www.sybex.com/go/roadwayessentials, where we'll post additional content and updates that supplement this book if the need arises.

CHAPTER 1

Getting Started

Autodesk® Roadway Design for InfraWorks 360™ is an add-on module that runs within Autodesk® InfraWorks 360™. It provides additional functionality for designing roads from an engineering perspective rather than the preliminary layout and visualization perspective that is provided by basic InfraWorks. It is a powerful tool for anyone laying out roads in InfraWorks who has an understanding of roadway engineering principles.

In this chapter, you'll learn to

▶ **Identify and understand the capabilities of Roadway Design for InfraWorks 360**

▶ **Navigate the user interface**

Understanding the Capabilities of Roadway Design for InfraWorks 360

Roadway Design for InfraWorks 360 extends the capabilities of InfraWorks 360 by offering advanced design and analysis tools for roadway design. The following sections break down these capabilities into several major areas.

Engineering Geometry

The roads you create with basic InfraWorks do not employ geometry that is consistent with civil engineering practices. These roads are considered to be "sketched" objects with no engineering design principles applied. With Roadway Design for InfraWorks 360, you will be creating roads that are defined by horizontal curves, spirals, vertical curves, and other characteristics unique to engineered roads. You will have access to many tools that allow you to view and edit these engineering-based geometric properties. These include specialized grips, additional asset cards, additional panels, and others. In Figure 1.1, you see the PI, PC, and PT gizmos (if you don't know what these abbreviations mean, refer to the "P-What?" sidebar), which indicate the geometry of an engineered horizontal curve, along with a Road asset card showing design speed, design standards, and other information. This is much more sophisticated than the sketched roads created by the basic InfraWorks tools.

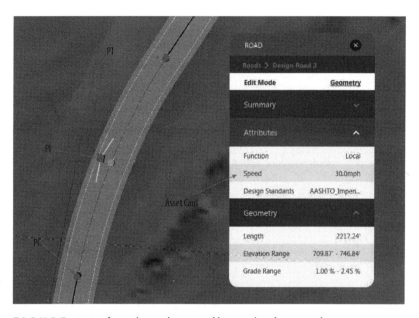

FIGURE 1.1   An engineered curve and its associated asset card

P-WHAT?

If you're new to the engineering aspects of road design, it might help to know what some of the common terms and abbreviations are. These terms and abbreviations can vary based on location. The ones listed here are commonly used in much of the United States.

Baseline   A *baseline* is a geometric pathway that the road geometry is based on. Often this is the centerline of the road, but it doesn't have to be.

Station   *Station* refers to your location along a baseline and is expressed using plus sign notation. If station 0+00 (0+000) is at the beginning of the baseline, then station 13+50 (1+350) would be 1,350 feet (meters) away from the beginning but along the path of the baseline.

(Continues)

P-WHAT? *(Continued)*

Offset   *Offset* is the perpendicular distance from the baseline to a given point. The locations of objects in reference to the baseline are often expressed using a station and offset.

Tangents (Baseline Segments)   *Tangents* are the straight-line portions of a baseline.

Tangent (Geometric Condition)   A *tangent (geometric condition)* is the kind of tangent you learned about in high school.

> ▶ It means touching or passing through at a single point.

> ▶ In the case of a line and arc, it means perpendicular to a line drawn from the intersection point to the center point of the arc.

> ▶ In the case of two arcs, it means intersecting in such a way that a line drawn from the center point of one arc to the center point of the other arc passes through the intersection point.

Curves   *Curves* are the curved portions of a baseline that have a constant radius.

Spiral   A *spiral* is a curved portion of a baseline that changes in radius from one end to the other. Most curves have a spiral placed at the beginning and end.

Point of intersection (PI)   PI is the place where two tangents intersect or would intersect if they were extended.

Point of curvature (PC)   PC is the place where the curve begins.

Point of tangency (PT)   PT is the place where the curve ends.

Point of reverse curvature (PRC)   PRC is **the** place where one curve meets another and the curve directions are different.

Point of compound curvature (PCC)   PCC is the place where one curve meets another and the curve directions are the same.

Rules-Based Design

When you create roads with the tools provided in Roadway Design for InfraWorks 360, design standards are built in. At the time that this book was written, the only standard available in the software was AASHTO 2011, but the availability of additional standards is expected. When you create a road, you

AASHTO stands for the American Association of State Highway and Transportation Officials.

choose from several classifications (see Figure 1.2). Based on the classification you choose, a design speed is assigned that then automatically determines horizontal and vertical geometry for your road. In other words, as you click points on the screen, the software is automatically designing the road for you.

FIGURE 1.2 The available road classification options

Profiles

If you have performed road design before, perhaps using AutoCAD Civil 3D, you know that the profile view is a critical tool in any road design. Roadway Design for InfraWorks 360 provides the Profile View panel (see Figure 1.3) where you can view your design and have a complete set of tools for modifying the design in profile view.

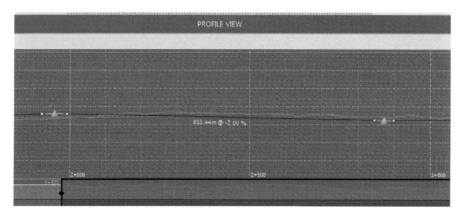

FIGURE 1.3 The Profile View panel

Intersections

Intersection design is often one of the most challenging aspects of any road design. In basic InfraWorks, the software handles road intersections to some extent, but there is a limited amount of control afforded to the user. Roadway Design for InfraWorks 360 enables design of the intersection through specialized gizmos and asset cards. Figure 1.4 shows a four-way intersection with the Intersection asset card. Notice the option where you can select a design vehicle and InfraWorks will automatically design the intersection geometry based on your choice.

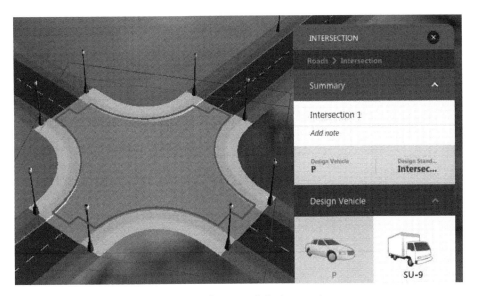

F I G U R E 1 . 4   Specialized gizmos and asset cards for intersections

Sight Distance Analysis

One of the most important aspects driving the design of a road is sight distance. With Roadway Design for InfraWorks 360, you can perform a sight distance analysis for a road that will assess the geometry of the road as well as components of the model that may serve as obstructions. The analysis provides useful graphical and textual feedback in the forms of colored bands, tooltips, and sight pins. Figure 1.5 shows a sight distance analysis that highlights an accident zone and an area of sight failure.

FIGURE 1.5   A sight distance analysis

Profile Optimization

Earthwork is another critical aspect in performing road design, one that dramatically affects cost. Roadway Design for InfraWorks 360 provides a powerful optimization tool that will analyze and adjust your road profile to minimize the costs associated with earthmoving. Because of the computing power required to perform an optimization, InfraWorks uses the cloud via InfraWorks 360 for this feature. Your design is uploaded to the cloud, analyzed, and adjusted, and a new profile is sent to you when the optimization is complete. This new profile is accompanied by a report describing the resulting geometry along with a detailed earthwork and cost analysis. Figure 1.6 shows the Profile Optimization panel where an optimization is configured.

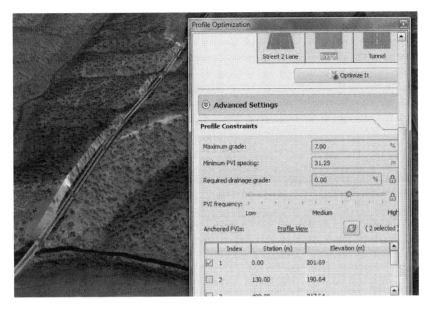

FIGURE 1.6 A profile optimization

Generate Civil 3D Drawings

When your InfraWorks design is complete and you're ready to move on to detailed design and documentation, Roadway Design for InfraWorks 360 provides a tool that will automate the process of converting your InfraWorks model into a series of DWG files that can be opened within Autodesk AutoCAD Civil 3D. Figure 1.7 shows a plan and profile drawing generated by the View Civil 3D Drawings command in InfraWorks. The command not only has created the necessary drawing files but also has created Civil 3D alignments and profiles and generated a series of plan and profile sheets complete with properly configured viewports.

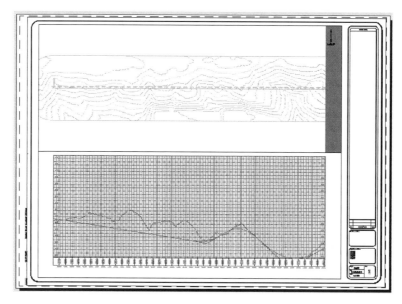

FIGURE 1.7   A Civil 3D plan and profile drawing generated by InfraWorks

Navigating Roadway Design for InfraWorks 360

The user interface for Roadway Design for InfraWorks 360 is provided as its own set of Intelligent Tools. The Home icon for these Intelligent Tools appears at the top left of the InfraWorks window as a tan circular icon with the tooltip Design, Review And Engineer Roads, as shown in Figure 1.8.

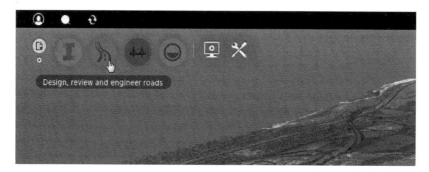

FIGURE 1.8   The Roadway Design Home icon

When you click the Roadway Design Home icon, it expands to reveal a toolbar containing three icons, as shown in Figure 1.9. I'll refer to this as the Roadway Design toolbar and the individual icons contained within it as the Analysis, Design, and Review icons.

FIGURE 1.9   The expanded view of the Roadway Design toolbar

Each of the icons on this toolbar opens another toolbar on the side of the InfraWorks window. The following sections cover the toolbars and the tools on them.

Analysis Toolbar

On this toolbar you will find tools for analyzing your model.

Terrain Themes   Terrain Themes is a basic InfraWorks command. It opens the Terrain Themes panel where you can configure a theme to color-code a terrain based on elevation, slope, or aspect (direction in which hillsides face).

Profile Optimization   Use this command to have InfraWorks 360 calculate the best profile design for your road based on cost, as well as design considerations such as maximum grade and PVI frequency.

Sun & Sky   Sun & Sky is a basic InfraWorks command. It opens the Sun & Sky asset card where you can adjust the appearance of light, shadow, and sky animation.

Design Toolbar

On this toolbar you will find tools for creating new design.

Highway Roads   This command launches road layout using the Highway design classification. This classification uses the highest design speed, resulting in larger horizontal curve radii, longer vertical curves, and other geometric properties consistent with high-speed roads. The design speed for the Highway classification is 70 mph (110 km/h).

Arterial Roads   This command launches road layout using the Arterial design classification. This classification is typically used for high-volume urban roads. The design speed is 50 mph (80 km/h).

Collector Roads   This command launches road layout using the Collector design classification. This classification is typically used for low-volume urban roads that provide access to residential areas. The design speed is 40 mph (60 km/h).

Local Roads   This command launches road layout using the Local Road design classification. This classification is typically used for low-volume urban roads that provide access to individual residential properties. The design speed is 30 mph (45 km/h).

City Furniture   This is a basic InfraWorks command. It allows you to add detail to your model by inserting 3D models of things such as light poles, cars, people, and a whole list of other items. Plus, you can include your own 3D models, so just about anything can be added to your model as city furniture.

Coverages   This is a basic InfraWorks function. You can think of a coverage as an area of land that is covered with something. That can be grass, pavement, concrete, sand, or just about anything. Coverages can also be used to shape the land that they cover.

Points Of Interest   This is a basic InfraWorks command. Create a point of interest to call out an important location in your model. Any 3D model can be used as a marker for a point of interest.

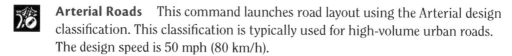 Review Toolbar

On this toolbar you will find tools for reviewing, analyzing, and exporting your design.

Profile View   This tool opens the Profile View panel where you can view and edit a road design in profile view. From here you can adjust PVI locations and elevations, add new PVIs, and edit vertical curves.

Sight Distance Analysis   This command opens the Sight Distance panel where you can configure and run a sight distance analysis. The analysis will tell you where there are sight distance failures and accident zones through a series of visual feedback and tooltips.

Surface Opacity   This is a basic InfraWorks feature. It is a handy toggle that switches your terrain surface from see-through to opaque. It is great for working on underground features such as pipes or bridge foundations.

Sun & Sky   Sun & Sky is a basic InfraWorks command. It opens the Sun & Sky asset card where you can adjust the appearance of light, shadow, and sky animation.

Profile Optimization   This command opens the Profile Optimization panel where you make choices and set values related to design constraints, quantities, cost, and construction rules. Once you've made your choices and run the optimization, pertinent data from the model is uploaded to the cloud via InfraWorks 360. Then, after processing the data, a revised profile and an optimization report are sent to you.

View Civil 3D Drawings   This command opens the Create Civil 3D Drawings dialog where you configure the options for creating drawings of your road design. From here you select the road you want to export, configure the area of the surface that will be included, and provide information that will configure the resulting drawing files and sheets.

Job Monitor   This command opens the Job Monitor panel where you can check on jobs that you have submitted using the Profile Optimization command. Here you can find out the status of the jobs and also download the resulting output when a job is complete.

Exercise 1.1: Explore Roadway Design for InfraWorks 360

To begin this exercise, go to the book's web page at www.sybex.com/go/roadwayessentials and download the files for Chapter 1. Unzip the files to the correct location on your hard drive according to the instructions in the introduction.

1. Launch InfraWorks 360.

2. On the Start Page, click Open.

3. Browse to C:\InfraWorks Roadway Essentials\Chapter 01\ and select Ch01 Bimsville Roads.sqlite. Click Open.

   You are looking at a model of the Bimsville Bypass project. If you have worked through the book *Autodesk InfraWorks and InfraWorks 360 Essentials*, then you are quite familiar with this project. In this version of the model, only a portion of the bypass has been created.

4. Click the Bookmarks icon on the Utility Bar; then click Bridge Plan to restore that bookmark.

   You are looking at a bridge in a top-down, or *plan view*, orientation.

5. Click the road to select it.

   The road should highlight in yellow, and the Road asset card should appear immediately, as shown in Figure 1.10, although the Summary and Attributes sections may not be expanded in your view.

FIGURE 1.10   A selected road and its associated asset card

6. With the road still selected, click Edit on the Utility Bar.

7. Zoom out so that you can see the gizmos associated with the curve to the northeast of the bridge.

   If you are familiar with the sketched roads created by basic InfraWorks, you should notice right away that the gizmos are differ-ent. Also notice that the asset card has an additional section entitled Geometry (see Figure 1.11).

If you do not see the Geometry section, select Geometry for Edit Mode.

8. Orbit the model so that you are now looking at the bridge from a lower angle. Continue to rotate the view until the gizmos change.

FIGURE 1.11　Specialized curve gizmos and a Geometry section on the Road asset card

Once you've rotated the view beyond a certain point, the gizmos change to those intended for vertical editing (see Figure 1.12). These gizmos demonstrate the advanced, engineering-based road geometry that is provided by the presence of Roadway Design for InfraWorks 360.

FIGURE 1.12　Specialized gizmos for vertical editing

9. Click the Roadway Design icon to reveal the Roadway Design toolbar.

10. On the Roadway Design toolbar, click the Analyze icon to open the Analyze toolbar along the left side of the InfraWorks window.

11. On the Roadway Design toolbar, click the Design icon to open the Design toolbar along the left side of the InfraWorks window.

12. On the Roadway Design toolbar, click the Review icon to open the Review toolbar along the left side of the InfraWorks window.

13. With the road still selected, click Profile View.

The Profile View panel should open showing you the design of the road in profile view, as shown in Figure 1.13. The light blue triangles are PVIs.

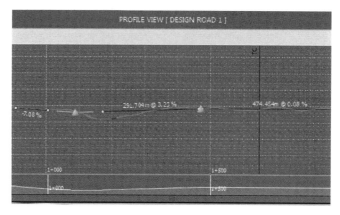

FIGURE 1.13 The Profile View panel

14. Click Close The Current Model in the upper-left corner of your InfraWorks screen.

The model is closed, and you are returned to the start page.

Now You Know

Now that you have completed this chapter, you understand the capabilities of Roadway Design for InfraWorks 360. You understand that it runs in InfraWorks 360 and extends its functionality by providing road design tools with engineers in mind. You are aware that these tools enable you to design and modify roads using engineering-based geometry such as horizontal curves, spirals, and vertical curves. You also know that you will have access to the profile view of your road designs—a critical requirement for performing road design effectively.

You now know that with the extended functionality provided by Roadway Design for InfraWorks 360, you can perform powerful functions such as automated intersection design, sight distance analysis, and profile optimization. You also have learned that with one command, you can generate a whole set of engineering drawing files that can be opened in Civil 3D where detailed design and documentation can be done.

Finally, after completing this chapter, you took a brief tour of the user interface that controls Roadway Design for InfraWorks 360, and you're ready to begin using it to perform some serious road design.

CHAPTER 2

Designing Roads

Now that you have familiarized yourself with Autodesk® Roadway Design for InfraWorks 360™, it's time to start putting the tools to use. As you have learned, these tools approach roadway design from an engineer's or designer's perspective, employing concepts that are consistent with the geometry and editing practices used for engineered roads. In this chapter, you will use those tools to begin building the Bimsville Bypass as well as the local roads that serve the industrial park.

In this chapter, you'll learn to

▶ **Create roads using the Intelligent Tools**

▶ **Understand and utilize various asset cards**

▶ **Edit roads graphically using gizmos**

▶ **Edit roads using context menus**

▶ **Divide roads into zones and vary the styles, lanes, and roadside grading**

▶ **View and edit a road in profile view**

▶ **Design and edit intersections**

Creating Roads

Of all the tasks that you'll perform with Roadway Design for InfraWorks 360, creating roads is probably the easiest. You will create roads using the Intelligent Tools on the Design toolbar. On the Design toolbar, you will find four Intelligent Tools for creating roads, each intended for a different function. They are Highway, Arterial, Collector, and Local (see Figure 2.1).

SPLINE ROADS VS. DESIGN ROADS

As discussed in Chapter 1, the roads you create with basic InfraWorks are different from those created with Roadway Design for InfraWorks 360. Spline roads are those originally created using basic InfraWorks and the sketch tools in the InfraWorks menu. Spline roads use geometry that looks good and is simple to work with when in planning and preliminary design stages. Design roads have more "bells and whistles" and take a bit more effort to construct—but for good reason. The geometry of design roads follows sound engineering principles and is more suited for engineering and eventually construction.

It is a common scenario to have one person sketch out the roads during planning and preliminary design and then pass that layout to someone else who "takes it to the next level" as far as design is concerned. It is also common for the person working in a planning capacity to use only basic InfraWorks, not Roadway Design for InfraWorks 360.

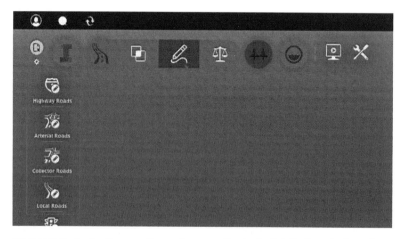

FIGURE 2.1   To access the roadway creation tools click the icons in the order shown

After clicking one of the roadway creation tools, the Select Draw Style asset card will open where you will select a style before drawing the road. When you click the first point of your new road, a tooltip will appear that displays design speed and length. This is your only opportunity to manually set the design speed. Once the road has been created, the design speed cannot be changed.

DESIGN ROADS ARE JUST AS STYLISH AS SPLINE ROADS

Even though design roads are fundamentally different from spline roads, they both use the same set of styles.

After that, it's just a matter of picking points within the model, each one becoming a PI for the road. As you draw, you'll notice that curves are created automatically at each PI (as long as there is room for a solution) and that a default design speed is automatically applied (see Figure 2.2). The geometry of the curves and other characteristics of the design are all happening automatically as you click each point.

FIGURE 2.2   Creating a new road

When you reach the final point on your road, you simply double-click to end the command. You will find that InfraWorks has made choices for you in regard to not only the horizontal layout but also the vertical design. You'll notice that there are also PVIs created at key changes in the existing terrain as the program tries to follow existing elevations while maintaining design standards (see Figure 2.3). It may not be the final design, but it is usually a great start that requires only a bit of tweaking.

FIGURE 2.3   The red arrows indicate locations where PVIs were created automatically.

Exercise 2.1: Create Design Roads

In this exercise you will create a portion of the Bimsville Bypass highway using the Intelligent Tools.

Go to the book's web page at www.sybex.com/go/roadwayessentials and download the files for Chapter 2. Unzip the files to the correct location on your hard drive according to the instructions in the introduction.

1. If it is not already open, launch InfraWorks 360. If you have a model open, close it to return to the InfraWorks Start Page.

2. On the start page, click Open, browse to C:\InfraWorks Roadway Essentials\Chapter 02\, and select Ch02 Bimsville Roads.sqlite. Click Open.

3. On the Utility Bar at the top right of your screen, click the Proposals drop-down list and select Ex_2_1.

4. On the Utility Bar, click the Bookmarks drop-down list and select Begin Road to restore that bookmark.

    You should see a large red pushpin marking the location of the first PI of the road section you're about to create.

5. Click the Roadway Design icon to expand the Roadway Design toolbar.

6. Click the Design icon to open the Design toolbar along the left side of your screen.

> The Utility Bar is the black strip at the top of your InfraWorks screen.

7. On the Design toolbar, click Highway Roads.
   The Select Draw Style asset card will open.

8. On the Select Draw Style asset card, click Change Display Options.

9. Click Show Content Items With Text And Thumbnail.

10. Click Street > Divided Highway.

11. Click the ground at the location of the pushpin marker.

12. Zoom out and pan northward. Click the next two pushpin markers, working northward.

13. Continue panning northward and double-click the final pushpin marker.
    After a pause, a new section of road will appear. It will be selected with the gizmos visible.

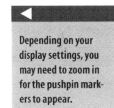

Depending on your display settings, you may need to zoom in for the pushpin markers to appear.

14. With the road still selected, restore the bookmark named Road View.
    You should be viewing the new road looking northward. Since the road is still selected, it should be highlighted in yellow, and the vertical design gizmos should be visible, as shown in Figure 2.4. You'll learn more about the gizmos later in this chapter.

FIGURE 2.4 A newly created road

You can view the results of completing the exercise successfully by choosing the Ex_2_1_End proposal.

Introducing the Road Asset Card

The Road asset card (pictured in Figure 2.5) is the primary editing tool in Roadway Design for InfraWorks 360. As you'll learn, it operates in several modes, each one providing a different set of tools. These tools can be used for graphical editing, changing styles, changing the number of lanes, and performing road-side grading. Certain edit modes allow the road to be subdivided into zones with different configurations applied differently in each zone. For edit modes that employ zones, a small flyout control provides tools for creating and removing zones (see Figure 2.5). A right-click menu also provides the same functions.

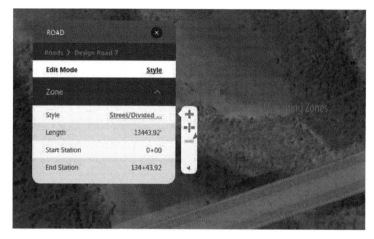

F I G U R E 2 . 5   The Road asset card shown with the control for managing zones

Editing Roads Graphically

In this section you'll study the use of the Geometry mode of the Road asset card. This is the edit mode that enables the editing gizmos as well as some specialized context menu commands. It also displays information about the road such as its function, design speed, and applied standards, as well as geometric properties such as overall length, elevation range, and grade range. The only values that can be edited in this mode are the name of the road and the note beneath it. Figure 2.6 shows the Road asset card in Geometry mode.

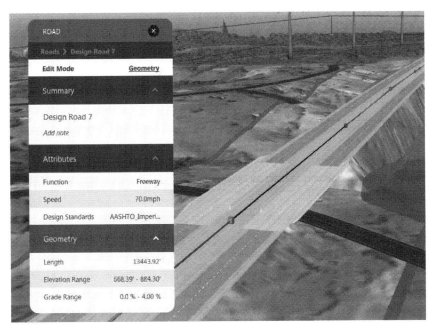

FIGURE 2.6 The Road asset card in Geometry mode with corresponding geometric editing gizmos

Editing with Gizmos

Two distinct sets of gizmos are available for editing roads: those for horizontal editing and those for vertical editing. The gizmos presented to you depend on your viewing angle. If you view the road in plan view (top-down), then you are presented with gizmos for horizontal editing. If you orbit the model so that you are viewing it from a lower angle (I'll refer to this as *3D view*), the vertical editing gizmos are presented.

Remember that you have to be in edit mode, and the Edit Mode setting on the Road asset card must be set to Geometry for the gizmos to appear. If edit mode is not activated, you can turn it on by clicking the Edit icon on the Utility Bar or by selecting the road, right-clicking, and selecting Edit. When you click one of the gizmos, you can change its location graphically by dragging and releasing in a different location, or you can enter values in the tooltip to change it numerically.

The following are the gizmos you have access to:

Horizontal Editing Gizmos   These gizmos are visible when you're viewing the model in plan view.

PI   Click and drag this gizmo to change the location of a PI.

Tangent   Click and drag this gizmo to move a tangent while maintaining its direction.

PC/PT   Click and drag these gizmos to change the length of the curve. These gizmos are not available for the spiral-curve-spiral curve type.

Radius   Click and drag this gizmo to change the radius of the curve. This gizmo is not available for the spiral-curve-spiral curve type.

Vertical Editing Gizmos   These gizmos are visible when you're viewing the model from a 3D view.

PVI   Click and drag this gizmo to change the elevation or station of a PVI. You cannot change the alignment of the road, only the location of the PVI along the alignment.

When you click a PVI grip, a cross-shaped graphic appears to show you that you can edit vertically or along the alignment. There are also light blue cone-shaped gizmos that enable you to move the PVI along the slope of the incoming or outgoing tangent. PVIs at the beginning or end of a road do not provide light blue cone-shaped gizmos. Figure 2.7 shows the cross-shaped graphic and cone-shaped gizmos. The light blue cylinder-shaped gizmo is discussed a little later.

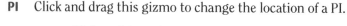

> ▶
>
> If you're not sure what a PI is, you should review the "P-What?" sidebar in Chapter 1.

F I G U R E 2 . 7   The editing gizmos located at the PVI of a road

PVC/PVT　　The purple cylinder gizmos are located at the PVC and PVT of a vertical curve. When you click and drag them, you change the length of the vertical curve. Dragging them away from the PVI makes the curve longer, and dragging them toward the PVI makes the curve shorter.

High/Low Point　　The light blue cylinder-shaped gizmo shows the location of a high point or low point for the vertical curve. It cannot be edited.

Exercise 2.2: Edit a Road Using Gizmos

In this exercise you will make some edits to the bypass road by utilizing several different types of gizmos. You will change the alignment of the road and change the elevations of the road so that there is enough vertical clearance to place a bridge over an existing road.

If you are continuing from the previous exercise, you can skip to step 3. Otherwise, if you haven't already done so, go to the book's web page at www.sybex.com/go/ roadwayessentials and download the files for Chapter 2. Unzip the files to the correct location on your hard drive according to the instructions in the introduction.

1. If it is not already open, launch InfraWorks 360.

2. On the Start Page, click Open and browse to C:\InfraWorks Roadway Essentials\Chapter 02\. Click Ch02 Bimsville Roads.sqlite and click Open.

3. Select the Ex_2_2 proposal. You should see the model as it would appear at the end of Exercise 2.1 with the new road section created.

4. Restore the bookmark named PI-1.

5. Click the road to select it. If the PI gizmo does not appear, right-click and select Edit. If you still do not see a PI gizmo, set Edit Mode to Geometry on the Road asset card.

6. Click and drag the PI gizmo to the center of the green pushpin marker. Release the gizmo when you are at the center of the pushpin marker.

7. Restore the bookmark named Tangent Gizmo.

8. Click and drag the light blue tangent gizmo to the center of the green pushpin marker. Release the gizmo when you are at the center of the pushpin marker.

9. Restore the bookmark named PVI-1. Click the PVI gizmo and then click the value for Elevation in the tooltip. Edit the value so that it reads 825 (250) and press Enter.

   After a pause, a portion of the road will move upward and should now appear as shown in Figure 2.8.

FIGURE 2.8   A view of the road after raising a PVI

10. Restore the PVI-2 bookmark. Click the PVI gizmo and change its elevation to 825 (250) in the same manner you changed the previous PVI.

11. Restore the bookmark named Bridge 2.

    As shown in Figure 2.9, with the last two edits, the new road is now elevated enough to create a bridge over the existing road.

FIGURE 2.9   Edits to the road have created enough vertical clearance for a future bridge.

You can view the results of completing the exercise successfully by choosing the Ex_2_2_End proposal.

Editing with the Context Menus

While in the edit mode of Geometry, a specific set of commands is available on a series of right-click context menus. The view direction (plan view vs. 3D view) and the location where you right-click will determine what commands are available. Let's break this down into four variations: plan view tangent, plan view curve, 3D view tangent, and 3D view vertical curve.

Plan View Tangent Commands   If you right-click on a tangent while in plan view, the menu will appear as shown in Figure 2.10. The commands included in it are as follows:

FIGURE 2.10   The road geometry context menu for tangents when viewing in plan view

Exit Edit Mode   Exits edit mode completely, enabling only a basic context menu.

Add Point   Adds a new PI at the location you've clicked.

Recreate Profile   Recalculates the PVIs and vertical curves based on the terrain. This will "reset" the profile and overwrite any manual edits you have made.

Add Bridge   Available only if Autodesk® Bridge Design for InfraWorks 360™ is enabled.

Add Culverts   Available only if Autodesk® Drainage Design for InfraWorks 360™ is enabled.

Properties   Opens the properties panel and displays the properties for the road.

Plan View Curve Commands   If you right-click a curve while in plan view, the following commands are available (see Figure 2.11):

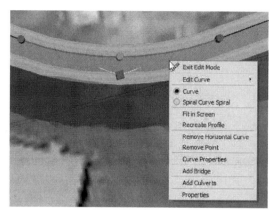

F I G U R E 2 . 1 1   The road geometry context menu for curves in plan view

Exit Edit Mode   This exits edit mode completely, enabling only a basic context menu.

Edit Curve   This tool becomes a flyout with sliders that control curve radius and spiral length (see Figure 2.12). Spiral length is available only for the spiral-curve-spiral curve type (covered in a bit). The low end of the sliders is unable to slide beyond the minimum values set forth by the currently selected design standards.

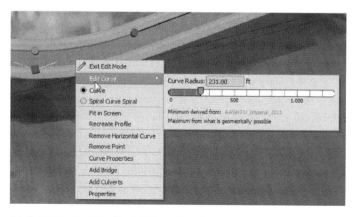

F I G U R E 2 . 1 2   The Edit Curve flyout

Curve   This configures the curve as a curve only (no spirals).

Spiral Curve Spiral   This configures the curve with a spiral at both ends.

Fit In Screen   This zooms your current view in to the extents of the curve.

Recreate Profile   This recalculates the PVIs and vertical curves based on the terrain. This will "reset" the profile and overwrite any manual edits you have made.

Remove Horizontal Curve   This removes the curve and any spirals, if applicable. There will still be a small curve, but you will not have the ability to edit it.

Remove Point   This removes the PI, thus straightening the road at that location.

Curve Properties   This opens the Curve Properties panel where you can view numerical information about the horizontal curve and associated geometry (see Figure 2.13).

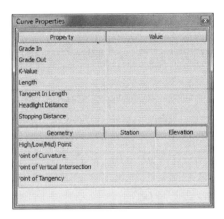

FIGURE 2.13 **The Curve Properties panel (shown undocked)**

Add Bridge   This is available only if Bridge Design for InfraWorks 360 is enabled.

Add Culverts   This is available only if Drainage Design for InfraWorks 360 is enabled.

Properties   This opens the properties panel and displays the properties for the road.

3D View Tangent Commands   If you right-click a tangent while in 3D view, the following commands are available (see Figure 2.14):

F I G U R E   2 . 1 4   The road geometry context menu for tangents in 3D view

Exit Edit Mode   Exits edit mode completely, enabling only a basic context menu.

Add PVI   Adds a new PVI at the location you right-click.

Recreate Profile   Recalculates the PVIs and vertical curves based on the terrain. This will "reset" the profile and overwrite any manual edits you have made

Add Bridge   Available only if Bridge Design for InfraWorks 360 is enabled.

Add Culverts   Available only if Drainage Design for InfraWorks 360 is enabled.

Properties   Opens the properties panel and displays the properties for the road.

3D View Vertical Curve Commands   If you right-click within a vertical curve while in 3D view, the following commands will be available (see Figure 2.15):

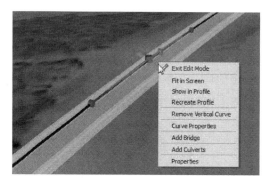

F I G U R E   2 . 1 5   The road geometry context menu for vertical curves in 3D view

Exit Edit Mode   Exits edit mode completely, enabling only a basic context menu.

Fit In Screen   Zooms to the extents of the vertical curve and orients the view to elevation view.

Show in Profile   Opens the Profile View panel (covered later in this chapter).

Recreate Profile   Recalculates the PVIs and vertical curves based on the terrain. This will "reset" the profile and overwrite any manual edits you have made.

Remove PVI   Deletes the PVI entirely, causing the vertical alignment to straighten in that area. This isn't shown because it is available only if you right-click one of the vertical curve gizmos.

Remove Vertical Curve   Deletes the vertical curve, leaving only an angle point at the PVI.

Curve Properties   Opens the Curve Properties panel where you can view numerical information about the vertical curve and associated geometry.

Add Bridge   Available only if Bridge Design for InfraWorks 360 is enabled.

Add Culverts   Available only if Drainage Design for InfraWorks 360 is enabled.

Properties   Opens the properties panel and displays the properties for the road.

Exercise 2.3: Edit a Road Using Context Menus

In this exercise you will use context menus to add a new section of road to the Bimsville Bypass.

If you are continuing from the previous exercise, you can skip to step 3. Otherwise, if you haven't already done so, go to the book's web page at www.sybex.com/go/roadwayessentials and download the files for Chapter 2.

Unzip the files to the correct location on your hard drive according to the instructions in the introduction.

1. If it is not already open, launch InfraWorks 360.

2. On the Start Page, click Open and browse to C:\InfraWorks Roadway Essentials\Chapter 02\. Click Ch02 Bimsville Roads.sqlite and click Open.

3. Select the Ex_2_3 proposal. Restore the bookmark named PI-3.

4. Click the bypass road to highlight it and show the Road asset card.

5. If the Edit Mode section is not visible on the Road asset card, right-click the road and select Edit. Set Edit Mode to Geometry.

6. Right-click the green pushpin located near the end of the road. Select Add Point.

   After a pause, a new PI will be created at the location of the green pushpin.

7. Click the purple cube gizmo at the end of the road and drag it to the other green pushpin to the north. Release it at the location of the pushpin.

   As you drag the end of the road to the new location, you can see the curve develop at the PI that you just created.

8. Right-click the new curve and select Spiral Curve Spiral.

   After a pause, you will see the curve broken into segments: two orange spiral segments and one blue curve segment.

9. Right-click the curve and move your cursor over the Edit Curve selection.

   The menu will expand to reveal sliders for the curve and spiral geometry.

10. Set Curve Radius to 2500 (765) and Spiral Length to 350 (100), as shown in Figure 2.16. Click a point in the model to exit the editing command.

    After a pause, the road model will be updated.

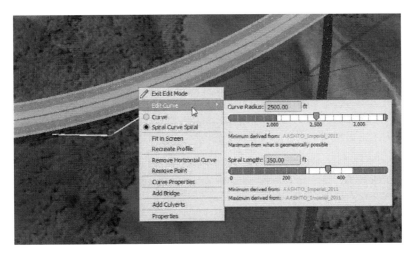

FIGURE 2.16 Setting the curve and spiral values

You can view the results of completing the exercise successfully by choosing the Ex_2_3_End proposal.

Working with Style Zones

In this section you'll be studying the use of the Style edit mode. With this edit mode you have your first introduction to zones. When you activate this mode, a blue outline is shown in the model to indicate which zone you are modifying. You can choose a different zone by simply clicking within it. Blue cylindrical gizmos also appear and can be clicked and dragged to graphically change the location of zone boundaries. The only editable property on the asset card for this mode is Style. This value is changed by clicking the style name to open the Select Style dialog and then choosing a style for that zone. Figure 2.17 shows the Road asset card in Style mode along with the blue zone outline, cylindrical zone boundary gizmos, and Select Style dialog. The flyout control is available in this mode for creating and removing zones.

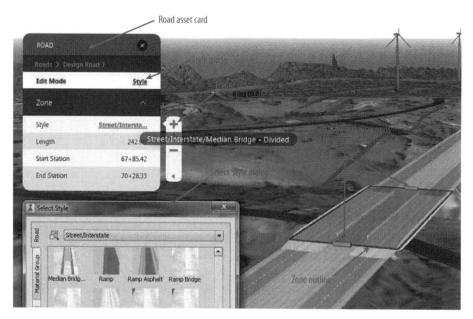

FIGURE 2.17   The Road asset card in Style mode with its corresponding functions

Exercise 2.4: Apply Style Zones

In this exercise you will create a new style zone in the location of a proposed bridge. You will apply a bridge style to that zone so that the design reflects the existence of the bridge.

If you are continuing from the previous exercise, you can skip to step 3. Otherwise, if you haven't already done so, go to the book's web page at www .sybex.com/go/roadwayessentials and download the files for Chapter 2. Unzip the files to the correct location on your hard drive according to the instructions in the introduction.

1. If it is not already open, launch InfraWorks 360.

2. On the Start Page, click Open and browse to C:\InfraWorks Roadway Essentials\Chapter 02\. Click Ch02 Bimsville Roads.sqlite and click Open.

3. Select the Ex_2_4 proposal. You should see the model as it would appear at the end of Exercise 2.3 with the edits made to the road.

4. Restore the bookmark named Bridge 2.

5. Click the bypass road to display the Road asset card. If the Edit Mode property is not visible, right-click the road and select Edit on the context menu.

6. Click the link next to Edit Mode and choose Style, as shown in Figure 2.18.

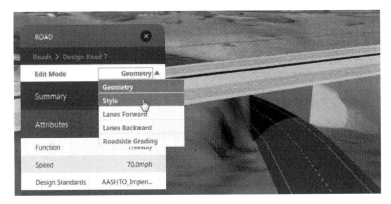

FIGURE 2.18   Choosing Style for Edit Mode

7. On the zone flyout, click Add A Style Zone, as shown in Figure 2.19. The Select Style dialog will open.

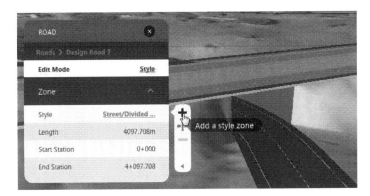

FIGURE 2.19   Splitting a zone

8. In the Select Style dialog, click the Road tab and choose the Street/ Interstate catalog from the drop-down list. Click Median Bridge - Divided and then click OK.

9. Move your cursor along the road, sliding the blue cylindrical gizmo as you do. Click when the station tooltip reads approximately 68+00 (2+075), as shown in Figure 2.20.

FIGURE 2.20   Setting the location between two zones

10. Move your cursor along the road and click a location near 69+50 (2+120) for the second gizmo. After a pause, the road style will change between the points you selected.

11. Press Esc to clear the selection. The bridge should now appear, as shown in Figure 2.21.

FIGURE 2.21   A bridge is represented using a style zone.

You can view the results of completing the exercise successfully by choosing the Ex_2_4_End proposal.

Hey, That's Not a Real Bridge

The bridge you created in Exercise 2.4 should not be confused with bridges created using Bridge Design for InfraWorks 360. Bridge Design for InfraWorks 360 is another module that can be enabled in InfraWorks 360. With this module you can employ engineering principles to design "real" bridges with control of the type, span, pier spacing, and many other options. Be sure to check the Wiley website to find out more about *Bridge Design for InfraWorks 360 Essentials*, an offering similar to this book, where you can learn how to use this module.

Working with Lane Zones

In this section you'll be studying the Lanes Forward and Lanes Backward edit modes. In these edit modes you are able to subdivide the lanes-forward or lanes-backward sides of the road into zones and change the number of lanes in each zone. This is handy for creating extra lanes in specific areas. As with other modes that involve zones, blue outlines show zone boundaries, and blue cylindrical gizmos enable the graphical editing of zone boundary locations. In Figure 2.22 you see the Road asset card in Lanes Forward mode along with the corresponding zone boundary and zone boundary gizmo. The flyout control is available in this mode for creating and removing zones.

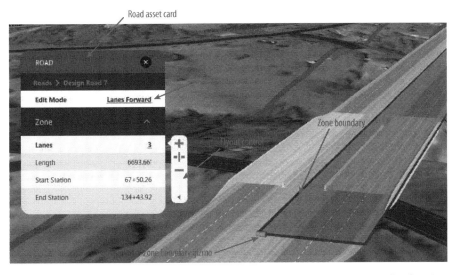

FIGURE 2.22 The Road asset card in Lanes Forward mode with its corresponding functions

Exercise 2.5: Edit the Number of Lanes

In this exercise you will use lane zones to create a deceleration lane leading up to a ramp that provides access to the industrial park.

If you are continuing from the previous exercise, you can skip to step 3. Otherwise, if you haven't already done so, go to the book's web page at www .sybex.com/go/roadwayessentials and download the files for Chapter 2. Unzip the files to the correct location on your hard drive according to the instructions in the introduction.

1. If it is not already open, launch InfraWorks 360.

2. On the Start Page, click Open and browse to C:\InfraWorks Roadway Essentials\Chapter 02\. Click Ch02 Bimsville Roads.sqlite and click Open.

3. Select the Ex_2_5 proposal.

    You should see the model as it would appear at the end of Exercise 2.4 with the bridge added near the industrial park. Also, a ramp and access road have been added that provide access to the industrial park from the bypass.

4. Restore the bookmark named Ramp 1.

    In this area you see the intersection of a ramp. You will be adding a deceleration lane to enable traffic to safely access the ramp.

5. Click the bypass road (outside the intersection area) to display the Road asset card.

6. If the Edit Mode section is not visible, right-click the road and select Edit.

7. On the Road asset card, change Edit Mode to Lanes Forward.

8. On the zones flyout, click Add A Lane Zone.

9. Pan southward and click the road near station 25+00 (0+765).

10. Pan northward and pick a point near station 45+00 (0+765).

11. On the Road asset card, change the value for Lanes to 3. After a pause to regenerate, an extra lane will be added in the zone you have selected.

12. Click the zone after the ramp and set the Lanes value to 2.

13. Press Esc to clear your selection.

14. Restore the bookmark named Decel Lane. Your view should look like Figure 2.23.

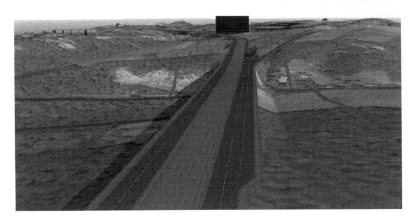

FIGURE 2.23   A deceleration lane created using the Road asset card with Edit Mode set to Lanes Forward

You can view the results of completing the exercise successfully by choosing the Ex_2_5_End proposal.

Configuring Roadside Grading

Roadside grading is the means by which InfraWorks calculates and builds the embankment from proposed road elevations to the terrain elevations. There are two methods used to calculate the locations where these embankments meet the terrain: Fixed Width and Fixed Slope.

Fixed Width

With Fixed Width, the location where the road embankment meets the terrain is at a constant distance from the edge of the road, while the slope varies at different locations along the road. The distance is set by the Grading Limit value (see Figure 2.24).

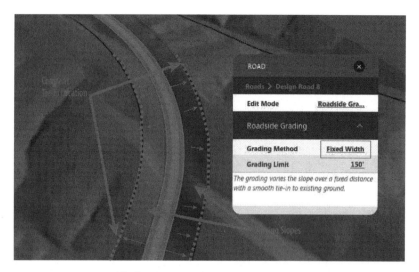

FIGURE 2.24   The Road asset card in Roadside Grading mode with Grading Method set to Fixed Width

Fixed Slope

With Fixed Slope, the slope is held constant, causing the location where the embankment meets the terrain to vary in relation to its distance from the road. If the end of the embankment lies outside the grading limits, then the embankment will end at the grading limits, and the grading will terminate by using a vertical face to intersect with the terrain. The slope itself is controlled by two values: Cut Slope and Fill Slope (see Figure 2.25).

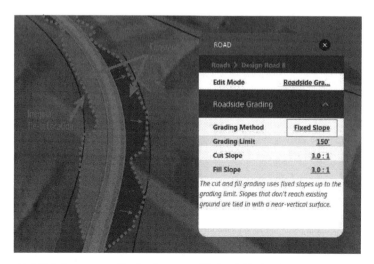

FIGURE 2.25   The Road asset card in Roadside Grading mode with Grading Method set to Fixed Slope

Exercise 2.6: Configure Roadside Grading

In this exercise you are going to change the roadside grading of the main road of the industrial park so that it is a fixed slope instead of fixed width. You will also study the effects of this change and the behavior of roadside grading in general.

If you are continuing from the previous exercise, you can skip to step 3. Otherwise, if you haven't already done so, go to the book's web page at www .sybex.com/go/roadwayessentials and download the files for Chapter 2. Unzip the files to the correct location on your hard drive according to the instructions in the introduction.

1. If it is not already open, launch InfraWorks 360.

2. On the Start Page, click Open and browse to C:\InfraWorks Roadway Essentials\Chapter 02\. Click Ch02 Bimsville Roads.sqlite and click Open.

3. Select the Ex_2_6 proposal.

   You should see the model as it would appear at the end of Exercise 2.5 with the deceleration lane and the access road to the industrial park.

4. Restore the bookmark named Industrial Park.

5. Click the main road running through the industrial park (located east of the buildings). When the Road asset card appears, if the Edit Mode section is not visible, right-click and select Edit.

6. On the Road asset card, change Edit Mode to Roadside Grading.

7. Restore the bookmark named Ravine 1.

   Notice how the embankment from the road to the terrain is very steep in the ravine area (see Figure 2.26). The grading method is currently set to the default, which is Fixed Width.

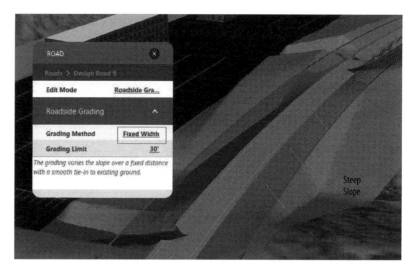

F I G U R E 2 . 2 6   Steep slope because of Grading Method set to Fixed Width

8. Change Grading Method to Fixed Slope.

After a pause, the roadside grading will change. Now there is a 3:1 slope projected from the edge of the road until it reaches the grading limit. After that, the embankment slope becomes a vertical face (see Figure 2.27).

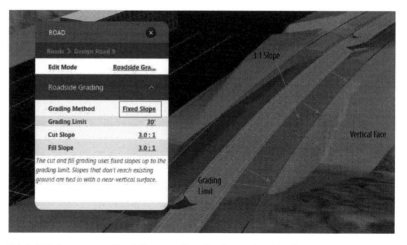

F I G U R E 2 : 2 7   Fixed Slope grading that is restricted by the Grading Limit setting

9. Change Grading Limit to 200 (60).

Since the embankment is able to meet the terrain before the Grading Limit setting, there is no vertical face needed (see Figure 2.28).

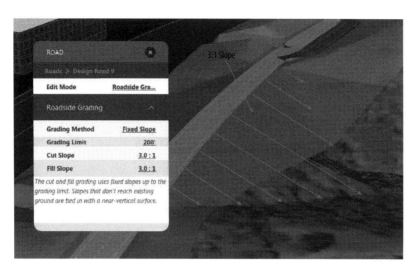

FIGURE 2.28 Fixed Slope grading that is unrestricted by the Grading Limit setting

You can view the results of completing the exercise successfully by choosing the Ex_2_6_End proposal.

Using the Profile View Panel

Profile view is an essential tool for anyone designing a road using engineering principles. In Roadway Design for InfraWorks 360, you are given easy access to the profile view of any design road through the Profile View panel. There are two ways to open the Profile View panel. The first is to click the Profile View icon on the Review toolbar, as shown in Figure 2.29.

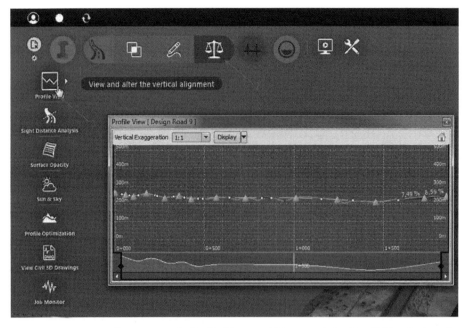

FIGURE 2.29 You can open the Profile View panel (shown undocked) by clicking the icons in the order shown.

PROFILE TERMINOLOGY

Familiarizing yourself with the following terms will be helpful as you work with design profiles:

Tangent   The straight-line portions of a profile

Point of vertical intersection (PVI)   The location where two tangents intersect

Point of vertical curvature (PVC)   The beginning of a vertical curve

Point of vertical tangency (PVT)   The end of a vertical curve

The other method is to right-click a road while in edit mode and while viewing in 3D view; then click the Show In Profile command, as shown in Figure 2.30.

FIGURE 2.30 Opening the Profile View panel using the right-click menu

The Profile View panel has several handy functions for changing the display and navigating the road profile (see Figure 2.31).

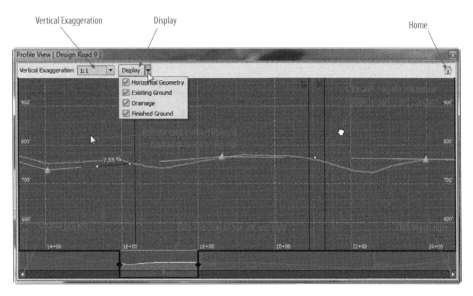

FIGURE 2.31 The Profile View panel

Vertical Exaggeration   In the top left is the Vertical Exaggeration setting where you can choose a ratio of vertical distances vs. horizontal distances.

Display   To the right of the Vertical Exaggeration feature is the Display setting where you can choose the components that are visible in your profile view. Your choices are Horizontal Geometry, Existing Ground, Drainage, and Finished Ground.

Home   The Home icon in the far-right corner will zoom out in profile view until you can see the entire profile. It behaves much like the Zoom Extents command in AutoCAD.

Pan and Zoom with the Mouse   Within the Profile View panel, you can use your mouse buttons in much the same way you navigate the 3D model. Click and drag the left button to pan, and roll the center wheel to zoom in and out.

Zoom by Station   Along the bottom of the Profile View panel you see two sets of stations listed and vertical bars that you can click and drag left and right. As you drag these bars inward, you are zooming in to a smaller area of the profile. The stations listed at the bottom represent the entire range of stations, and the stations above show you the station range that you are zoomed in to.

Pan by Station   Beneath the stationing is a horizontal scroll bar that you can move left and right by clicking and dragging or by using the arrow buttons at either end. As you move this bar, the area that you are zoomed into slides along the overall stationing, allowing you to move laterally along the profile view while maintaining your current zoom level.

It's OK to Exaggerate Sometimes

Vertical exaggeration is a common practice used to display information in profile view. In most places, the earth is relatively flat, and the changes in elevation are quite subtle. To make the peaks and valleys stand out a bit more, the elevations are exaggerated while the horizontal distances are kept the same. The result is a profile that appears to have higher high points, lower low points, and steeper slopes in between. This makes the terrain easier to visualize and analyze in profile view. For example, with a vertical exaggeration of 5:1, the height of peaks and the depth of valleys appear five times larger than the horizontal measurements.

Editing in Profile View

Once you have the Profile View panel open, you can edit the design profile (also called a *vertical alignment*) in a number of ways. One way is to click and drag a triangular PVI symbol to a new location. This is a powerful method for establishing the relationship between existing ground and proposed elevations (see Figure 2.32).

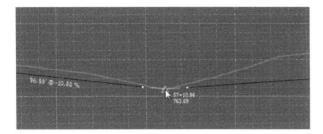

F I G U R E   2 . 3 2   Moving a PVI in the Profile View panel

In addition to graphically changing the profile there are a number of right-click context menu commands that are useful. Two context menus are available; one appears when you have selected a PVI, and the other appears when you have not.

Right-Click Menu: PVI Not Selected   If you right-click in the Profile View panel without first clicking a PVI, the shortcut menu will offer the following commands:

> **Add PVI**   Creates a new PVI in the location you right-clicked.
>
> **Recreate Profile**   Recalculates the placement and geometry of PVIs based on the location of the road in relation to the terrain. This will "reset" the profile and overwrite any manual edits you have made.
>
> **Properties**   Opens the properties panel for the road you have selected.

Right-Click Menu: PVI Selected   If you right-click in the Profile View panel after clicking a PVI, the shortcut menu will offer the following commands:

> **Show In Canvas**   Zooms in to the location of the PVI in model view.

WHAT'S A CANVAS?

Canvas is another word for the model. You'll see some tools and functions referred to as in-canvas, meaning they are manipulated right in the model area. In this book I usually refer to the canvas as simply the model, but it should be noted that *canvas* is Autodesk's official term for this.

Curve Properties   Opens the Curve Properties panel that displays information about the geometry of the vertical curve. This information cannot be edited.

Remove PVI   Deletes the PVI, thus straightening the profile at that location.

Remove Vertical Curve   Removes the curve, leaving only a grade break.

Recreate Profile   Recalculates the placement and geometry of PVIs based on the location of the road in relation to the terrain. This will "reset" the profile and overwrite any manual edits you have made.

Properties   Opens the properties panel for the road you have selected.

You should know that the road must be in edit mode for any of these functions to work.

Exercise 2.7: Edit in Profile View

In this exercise you are going to use the Profile View panel to make some changes to the vertical alignment of the industrial park road.

If you are continuing from the previous exercise, you can skip to step 3. Otherwise, if you haven't already done so, go to the book's web page at www .sybex.com/go/roadwayessentials and download the files for Chapter 2. Unzip the files to the correct location on your hard drive according to the instructions in the introduction.

1. If it is not already open, launch InfraWorks 360.

2. On the Start Page, click Open and browse to C:\InfraWorks Roadway Essentials\Chapter 02\. Click Ch02 Bimsville Roads.sqlite and click Open.

3. Select the Ex_2_7 proposal.

4. Restore the bookmark named Inlets.
   You are looking at a close-up view of the two inlets near the entrance to the industrial park. The inlets are projecting above grade.

5. Click the Roadway Design icon to open the Roadway Design toolbar.

6. Click the Review And Modify icon to open the Review And Modify toolbar along the left side of the screen.

7. On the Review And Modify toolbar, click Profile View.
   The Profile View panel will open, but it will be blank.

8. Click the industrial park road. If the Road asset card does not show the Edit Mode section, right-click and select Edit from the shortcut menu. For Edit Mode select Geometry.
   The profile of the road will appear in the Profile View panel.

9. On the Profile View Panel, use your mouse to zoom and pan so that you are focusing on the first three PVIs.

10. Right-click the second one and click Remove PVI, as shown in Figure 2.33.

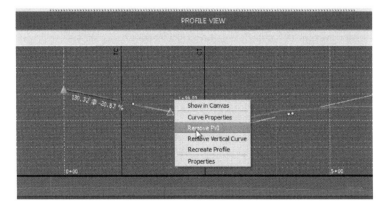

FIGURE 2.33   Removing a PVI in profile view

After a pause, the model and profile view will update. This PVI is unnecessary since the road profile is nearly straight in this area.

11. Click the PVI at the low point near station 3+50 (0+100) and drag it to the left. Release it when the tooltip reads an approximate station of 3+10 (0+094.50) and elevation of 795.50 (242.50).

As you drag the PVI in the Profile View panel, notice how the graphics change in the model view. When you release it, the road will rebuild, and now the inlets are flush with the road surface. Also notice that the low point of the curve is located between the inlets (see Figure 2.34).

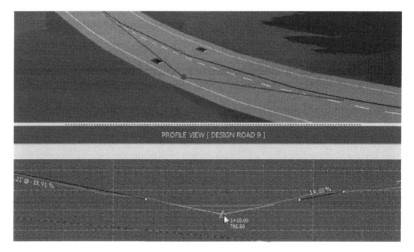

FIGURE 2.34 Editing a profile in the Profile View panel

You can view the results of completing the exercise successfully by choosing the Ex_2_7_End proposal.

Working with Intersections

In InfraWorks you don't even need to create intersections; they create themselves whenever two roads are in close enough proximity with one another. In basic InfraWorks, when dealing with spline roads, the intersection geometry is calculated by InfraWorks and cannot be modified by you, the user. However, with Roadway Design for InfraWorks 360 and the use of design roads, InfraWorks does calculate the initial design of the intersection, but then it gives you tools to refine that design to suit your needs.

The Intersection Asset Card

When you click an intersection, the Intersection asset card will open. If you are not in edit mode, you will see only the Summary section, which reports the name and note for the intersection along with the design vehicle and intersection standard. None of these values can be edited if you are not in edit mode. With edit mode turned on, you can edit the name and note as well as select a design vehicle for the intersection (see Figure 2.35). This is if the entire intersection is selected, not an individual zone (I'll discuss zones in just a bit). The design vehicle that you select will dictate the default values for curb return radii, taper lengths, and other intersection geometry based on the intersection standard that has been selected. Generally speaking, larger, less agile vehicles will require larger radii and larger dimensions overall.

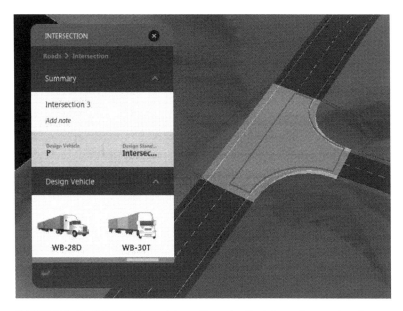

FIGURE 2.35 Editing an intersection using the Intersection asset card

The Turning Zone Asset Card

Intersections are automatically subdivided into zones: four zones for a four-way intersection and two zones for a three-way intersection. After you initially select an intersection, if you click again to select one of the zones, the Turning Zone asset card will open (see Figure 2.36). With this asset card you can control the curb-return curve type (Simple Curve or Simple Curve With Taper). You can also control the values that configure the geometry of the curve type you've selected.

FIGURE 2.36   Editing a turning zone

Simple Curve   For a simple curve, the curb return is created as a simple fillet using a single curve.

Simple Curve with Taper   For this configuration, the lane is widened outward by a distance equal to the Offset value. This is applied to the incoming and outgoing road. The Taper value controls the angle of the taper and consequently the length of the transition area between the normal road width and the application of the offset; the smaller the value, the more abrupt the taper. So, for example, a road that has an offset of 5′ (2 m) and a taper of 5:1 will require a transition area length of 25′ (10m) (see Figure 2.37).

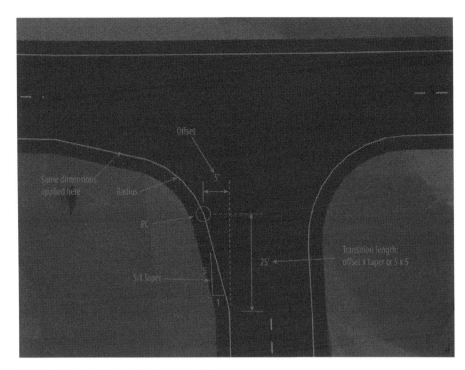

FIGURE 2.37 A simple curve with taper

Radius   This is the radius of the curve that provides the fillet between the sides of the intersecting roads. This applies to the Simple Curve and Simple Curve With Taper configurations.

Offset   This is the amount of widening applied to the lane as it progresses toward the intersection.

Taper   This is the ratio of length to width that provides the transition from the normal road width to the application of the offset.

Intersection Gizmos

As with most items in InfraWorks, intersections provide a set of specialized gizmos to make editing easy and visual. There are two types of gizmos: the light blue cone gizmo that is used to control curb return radius, and the purple cylinder gizmo that is used to add a widening zone. For the light blue cone gizmo, you simply click and drag—outward to increase the radius and inward to decrease it (see Figure 2.38).

FIGURE 2.38 The light blue cone gizmo (shown in red because it is selected) is used to control the curb return radius.

When you click and drag the purple cylinder gizmo away from the intersection, it creates a new type of zone called a *widening zone* (see Figure 2.39). It will add a lane to the road, configure the transition area for the widening zone, and recalculate the curb return so that the geometry is still correct. Most importantly, it creates a whole new component to the intersection that can be configured and edited. This is covered in the next section.

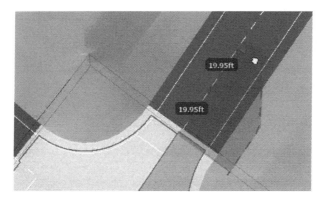

FIGURE 2.39 The purple cylinder gizmo (shown in red because it is selected) is used to add a widening zone.

Working with Widening Zones

Once you have created a widening zone for an intersection by clicking the purple cylinder gizmo, you are given access to a new type of asset card: the Widening asset card (see Figure 2.40). This asset card has the following functions:

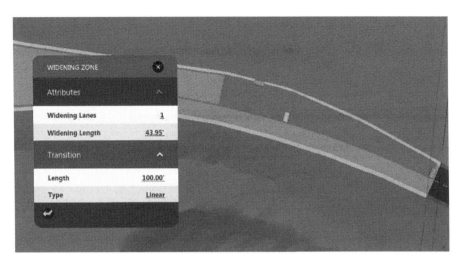

FIGURE 2.40 The Widening Zone asset card

Attributes These values are shown in the Attributes section of the Widening Zone asset card.

> **Widening Lanes** The number of additional lanes that are added to the original road
>
> **Widening Length** The length of the new lane(s)

Transition These values are shown in the Transition section of the Widening Zone asset card. They address the transition area between the addition of one or more lanes and the normal width of the road.

> **Length** This is the length of the transition from normal road width to the place where the additional lane (or lanes) starts.
>
> **Type** There are four choices for the geometric configuration for the transition area: Linear, Curve-Tangent-Curve, Curve-Curve-Reverse Curve, and Curve-Reverse Curve.

Curve 1 Radius   This is the radius of the curve adjacent to the additional lane(s). This option is not available when Type is set to Linear.

Curve 2 Radius   This is the radius of the curve adjacent to the normal road width. This option is not available when Type is set to Linear or Curve-Reverse Curve.

Reverse Curve Radius   This is the radius of the reverse curve adjacent to Curve 1. This option is not available when Type is set to Linear or Curve-Tangent-Curve.

A WAY BACK

You may have noticed the arrow icon at the bottom left corner of each of the asset cards associated with intersections. This icon will get you back to an original state which will vary based on which asset card you're dealing with. For the Intersection asset card, this will undo any changes that you've made after first selecting a design vehicle. This includes graphical edits with gizmos or any changes to the numerical design values on the asset card itself. For the Turning Zone asset card, the behavior is similar to that of the Intersection asset card, but the reset is limited to just the turning zone you've selected. And finally, for the Widening Zone asset card, the widening zone is removed completely by resetting the number of widening lanes to zero.

There are also gizmos associated with widening zones (see Figure 2.41). A light blue cylinder gizmo along the side of the added lane can be dragged outward to create additional lanes. A yellow cylinder gizmo can be dragged to set the length of the new lanes. And finally, a blue cylinder at the end of the transition area can be dragged to set the location of the end of the transition, which is also the end of the entire widening zone.

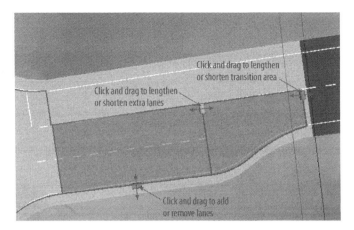

FIGURE 2.41 The widening zone gizmos

Exercise 2.8: Design an Intersection

In this exercise you are going to configure an intersection within the industrial park. You will design the intersection to accompany large tractor-trailers, and you will provide a turning lane to help with the flow of traffic.

If you are continuing from the previous exercise, you can skip to step 3. Otherwise, if you haven't already done so, go to the book's web page at www .sybex.com/go/roadwayessentials and download the files for Chapter 2. Unzip the files to the correct location on your hard drive according to the instructions in the introduction.

1. If it is not already open, launch InfraWorks 360.

2. On the Start Page, click Open and browse to C:\InfraWorks Roadway Essentials\Chapter 02\. Click Ch02 Bimsville Roads.sqlite and click Open.

3. Select the Ex_2_8 proposal.

4. Restore the bookmark named Intersection.

5. Click the intersection to open the Intersection asset card. If the Design Vehicle section of the Intersection asset card is not visible, right-click and select Edit.

6. For Design Vehicle, scroll to the right and click WB-67 (WB-20 for metric users).

   After a pause, the intersection geometry will change dramatically. This type of truck will need to be accommodated for the intersection area.

7. Click the northern turning zone to select it. Click the purple cylinder gizmo and drag it to the right as shown in Figure 2.42. Release when the red dotted outline of the widening zone appears.

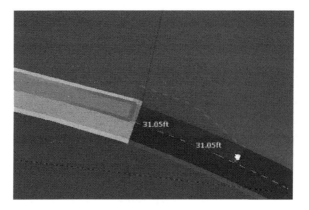

FIGURE 2.42   Creating a widening zone

8. Click the widening zone. On the Widening Zone asset card, do the following:

   ▶ Under Attributes, set Widening Length to 100 (30).

   ▶ Under Transition, set Length to 100 (30).

   ▶ Under Transition, verify that Type is set to Linear.

9. Press Esc to clear your selection.

   The intersection should look like Figure 2.43. This is a much more sophisticated intersection, and it allows for a left turning lane to help with traffic congestion at the intersection.

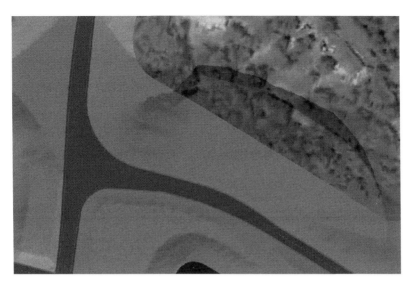

FIGURE 2.43   A designed intersection

Now You Know

Now that you have completed this chapter, you can create design roads by using the Intelligent Tools. You are able to edit roads using gizmos, context menus, and the Profile View panel. You are able to break a design road into zones and at different locations vary the style, the number of lanes, or the roadside grading configuration. Finally, you are able to create and modify intersections, including turn zones and widening zones.

Now that you have completed this chapter, you can begin creating and modifying design roads using Roadway Design for InfraWorks 360.

CHAPTER 3

Using Advanced Functions

Now that you have completed the initial design, you are ready to take steps to make it even better. Autodesk® Roadway Design for InfraWorks 360™ provides tools for optimizing and analyzing your design as well as a tool for pushing it into the detailed design stages that are taken care of by Autodesk® AutoCAD® Civil 3D®. By using these tools, you are moving your design from an early conceptual version toward a well-engineered version that is poised for detailed design.

In this chapter, you'll learn to

▶ **Balance earthwork and minimize cost using Profile Optimization**

▶ **Analyze roadways and intersections for sight distance issues**

▶ **Generate Civil 3D drawings of your road design**

Using Profile Optimization

When you first draw a road in InfraWorks using the Roadway Design for InfraWorks 360 tools, the software makes its best guess at balancing "smooth" road geometry with a close match to the existing terrain. As a result, you are given an initial design profile that is then modified and adjusted repeatedly as the design progresses. This profile is the "backbone" of the road model and is the single greatest influence on the earthwork quantities that will be produced by your road.

Questions about cut and fill quantities will be among the first questions posed about your design because excavation is such an expensive part of construction. As a designer, you will be called upon to minimize cut and fill quantities and to report your results in detail. Typically, adjusting the profile to

achieve the best earthwork results is an iterative process—a little higher here, a little lower there, and run the volume analysis. Then repeat as many times as necessary until earthwork results appear to be the best they can be. As you might guess,
or as you may have experienced, this process is time-consuming and repetitive.

What if you could click a magic button that would perform all of these iterations
for you? What if the computing power needed to perform such an optimization was
readily available? Well, through the cloud via InfraWorks 360 and a function called
Profile Optimization, this is all a reality. And not only will Profile Optimization balance your road earthwork, it will also provide a detailed cost analysis that you can
use to assess the overall cost of the road, not just earthworks.

Using the Profile Optimization Panel

To begin a profile optimization, you will need to open the Profile Optimization
panel. This is done by clicking the Analysis icon of the Roadway Design toolbar
and then clicking Profile Optimization on the Analysis toolbar (see Figure 3.1).

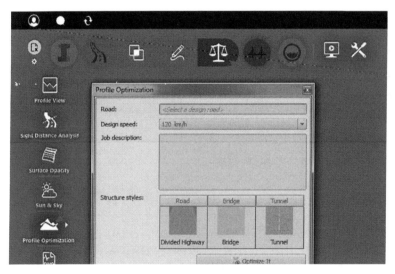

FIGURE 3.1   The Profile Optimization panel can be opened by clicking the
icons in the order shown.

In the top portion of the Profile Optimization dialog are some general settings
(see Figure 3.1). They are described as follows:

Road   This is the name of the road as you would see it in the Summary section
of the Road asset card.

Design Speed   The Profile Optimization function is going to redesign your profile. This setting will establish the design constraints for things such as vertical curve length, stopping sight distance, and so on.

Job Description   When you submit your optimization, it will be sent to InfraWorks 360 in the cloud as an optimization job, and you will be able to check on it using the Job Monitor (covered a little later in this chapter). Job Description will help you distinguish between multiple jobs if you've submitted multiple optimizations.

Structure Styles   When the optimization encounters things such as ravines and mountains, it will determine the most cost-effect course to take. Sometimes that will require tunneling through a mountain or placing a bridge across a ravine. This section is your opportunity to assign styles to those structures should they be needed.

Once you have selected a road for optimization, you will be given access to the Advanced Settings section of the Profile Optimization panel. The Advanced Settings section is broken into several subsections, with each containing a set of functions and options. A summary of each subsection and the items within it follows:

Profile Constraints   This section gives you the opportunity to establish some ground rules about how you want InfraWorks to redesign your profile (see Figure 3.2).

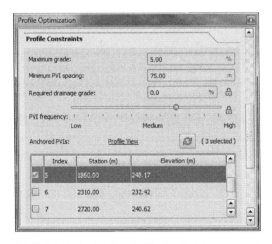

FIGURE 3.2   The Profile Constraints section of the Profile Optimization panel

Maximum Grade   This establishes the steepest grade allowable on the tangent sections of the profile. The initial value is set when you provide a value for Design Speed but you can change it after doing so.

Minimum PVI Spacing   The optimization process will create additional PVIs in your profile. This value establishes a minimum distance between them. Increasing this value will result in a "smoother" profile with fewer PVIs.

Required Drainage Grade   This is the minimum grade allowed for the tangent sections of the profile. Try to set this at the minimum value possible because if it is too high, it may prevent the optimization from finding a solution.

PVI Frequency   This will affect the "smoothness" as well but is a bit different from Minimum PVI Spacing. This is a more general setting that will affect the distribution of PVIs across the entire profile.

Anchored PVIs   In addition to the profile constraints just mentioned, there may be other considerations that require the profile to maintain certain elevations in a given area. This is your opportunity to "lock down" those areas by choosing PVIs that are to be preserved during the optimization. This means that during the optimization process they will not be deleted and they will hold their current station and elevation. For example, you might lock down PVIs near an existing road intersection or in front of a building where you know that the current road elevations must be preserved. The first and last PVIs are always anchored and cannot be changed.

You can click Profile View to open the Profile View panel while you are configuring Anchored PVIs. Also, if you double-click a PVI in the list, the model will zoom to that PVI, and it will be highlighted in red in the Profile View panel.

Quantities Options   In this section you can add borrow pits and waste pits to your design (see Figure 3.3). Each one is assigned Station, Access Distance, and Capacity. These locations will be considered in the mass haul aspects of optimizing the profile.

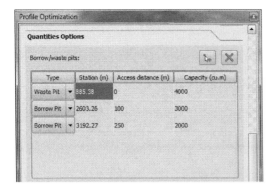

FIGURE 3.3 The Quantities Options section of the Profile Optimization panel

WHAT IS MASS HAUL?

Without getting too technical, *mass haul* is the concept of tracking the excavation, movement, and placement of earth during the process of constructing a road. By analyzing mass haul, these activities can be minimized, thus minimizing the cost to construct the road.

Construction Rules   In this section, you specify the thresholds for the creation of bridges and tunnels (see Figure 3.4). For bridges, this is a maximum fill height, and for tunnels this is a maximum cut depth.

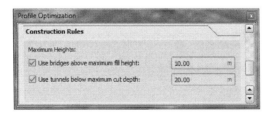

FIGURE 3.4 The Construction Rules section of the Profile Optimization panel

Construction & Earthwork Cost   In this section you assign cost values for earthwork and construction. Earthwork costs include items such as excavation, borrow, waste, and so on, while construction costs include items for pavement, bridges, tunnels, signing, and others. To access these settings, you must sign in to InfraWorks 360 and then click the link in the Profile Optimization dialog to open the Construction & Earthwork Costs Settings dialog (see Figure 3.5).

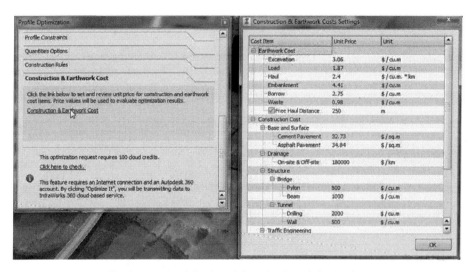

FIGURE 3.5   The Construction & Earthwork Costs Settings dialog has been opened by clicking the link in the Profile Optimization dialog.

Once you have addressed everything you need in the Profile Optimization panel, you click Optimize It (see Figure 3.6). When you do, InfraWorks will package all of the information it needs and submit it to the cloud for analysis.

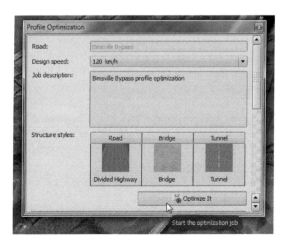

FIGURE 3.6   Launching a profile optimization

Exercise 3.1: Optimize Your Profile

In this exercise you will optimize the profile for a section of the Bimsville Bypass road.

ATTENTION: THIS EXERCISE REQUIRES CLOUD CREDITS

You are going to perform a function during this exercise that requires cloud credits. Please check with the person in your company or organization who is responsible for issuing cloud credits to ensure that you are permitted to use them for this purpose.

If you are unfamiliar with the cloud credit concept, you can think of them as Autodesk 360 currency. Your company will buy cloud credits that you can use to perform functions such as the profile optimization I'm discussing here. They cost money, so be careful with how you use them and get permission before you do.

If you do not have access to cloud credits, pay attention to the exercise instructions, and you will be notified about which steps you cannot perform.

Go to the book's web page at www.sybex.com/go/roadwayessentials and download the files for Chapter 3. Unzip the files to the correct location on your hard drive according to the instructions in the introduction.

1. If it is not already open, launch InfraWorks 360. If you have a model open, close it to return to the InfraWorks Start Page.

2. On the start page, click Open, browse to C:\InfraWorks Roadway Essentials\Chapter 03\, and select Ch03 Bimsville Roads.sqlite. Click Open.

3. On the Utility Bar at the top right of your screen, click the Proposals drop-down list and select Ex_3_1_and_3_2.

4. Restore the bookmark named Bimsville Bypass.

5. Click the Roadway icon to expand the Roadway toolbar.

6. On the Roadway toolbar, click the Analysis icon to display the Analysis toolbar along the left side of the InfraWorks window.

7. On the Analysis toolbar, click Profile Optimization.
   The Profile Optimization panel will open.

8. Click the Bimsville Bypass road.

   When you select the road, the Road property on the Profile Optimization panel will automatically say Bimsville Bypass, and the Optimize It button will activate.

9. Click Advanced Settings and then click Profile Constraints.

10. Next to Anchored PVIs, click Profile View.

    The section of road that you are viewing has all of the PVIs removed, except for the first and last (see Figure 3.7). This was done so that the effects of the optimization are quite apparent when the optimization is complete. On the Profile Optimization panel you may notice that there are three PVIs. PVI 2 was created automatically because of an intersection with an existing road.

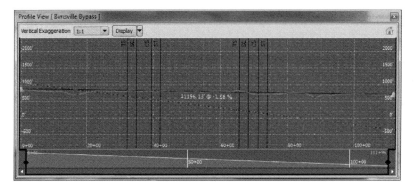

F I G U R E   3 . 7   The initial profile, prior to optimization

11. In the Profile Constraints section, under Anchored PVIs, uncheck the box next to PVI 2 so that the PVI is not anchored during optimization.

    By unchecking this box, you are enabling InfraWorks to "break" this intersection, allowing the new profile to pass above or below the existing road.

12. Also under Profile Constraints, do the following:

    ▶ For Minimum PVI Spacing, enter 500 (150).

    ▶ For Required Drainage Grade, click the lock icon to unlock the value. Enter 0.5 (for both Imperial and metric users).

    ▶ For PVI Frequency, click the lock icon to unlock the value and move the slider to the Medium position.

13. Click Optimize It.

    If you are not logged in or do not have cloud credits available, you will not be able to perform this step or the next step.

    If you are successful, a dialog will open indicating that the optimization is being processed. Then, after about 30 seconds, another dialog will open indicating that the data package was sent out for profile optimization.

14. Click OK to dismiss the Information dialog.

    The Job Monitor panel will open, and your recently submitted job should be displayed with a status of In Progress as indicated by the double blue arrows.

Getting Your Optimization Results

Once you have submitted a job for optimization, the Job Monitor panel will open, showing you all of the optimization jobs that you have submitted to InfraWorks 360. If you close this panel, you can open it by clicking Job Monitor on the Analysis toolbar (see Figure 3.8).

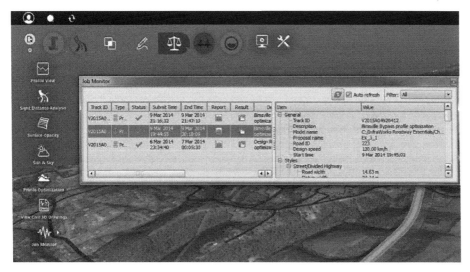

You will also receive an email informing you that optimization is complete.

F I G U R E 3 . 8   The Job Monitor panel (to open it, click the icons in the order shown)

When your optimization is complete, the Status column will display a green check mark. Also, when your optimization is complete, you will see icons in the Report and Result columns. If you click the icon in the Report column, you will download and open a PDF showing information about your optimization including the new profile design, cost of construction, earthworks data, and cross section views (see Figure 3.9).

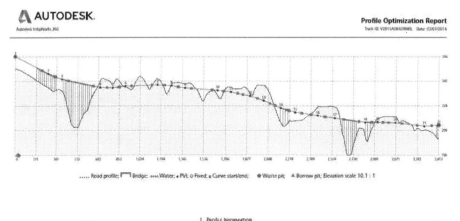

FIGURE 3.9 The first section of a Profile Optimization report

If you click the Results icon, you will download an IMX file and then be prompted to import the results into your InfraWorks model as a new proposal (see Figure 3.10). If you are satisfied with the results, you can use the Merge proposals function on the Proposals panel to replace your old road design with the optimized one.

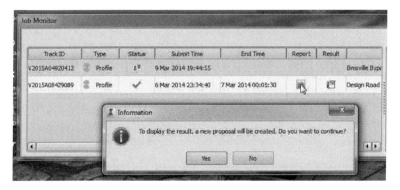

FIGURE 3.10 Importing optimization results into an InfraWorks model

Exercise 3.2: Get Your Optimization Results

In this exercise you will view the results of the optimization that you performed in Exercise 3.1. You will also update your model by creating a proposal from the optimized road and merging it with your current proposal.

If you are continuing from the previous exercise, you can skip to step 3. Otherwise, if you haven't already done so, go to the book's web page at www .sybex.com/go/roadwayessentials and download the files for Chapter 3. Unzip the files to the correct location on your hard drive according to the instructions in the introduction.

1. If it is not already open, launch InfraWorks 360.

2. On the Start Page, click Open and browse to `C:\InfraWorks Roadway Essentials\Chapter 03\`. Click `Ch03 Bimsville Roads.sqlite` and click Open.

3. On the Utility Bar at the top right of your screen, click the Proposals drop-down list and select Ex_3_1_and_3_2.

4. Restore the bookmark named Bimsville Bypass.

5. Click the Roadway icon to expand the Roadway toolbar.

6. On the Roadway toolbar, click the Analysis icon to display the Analysis toolbar along the left side of the InfraWorks window.

7. On the Analysis toolbar, click Job Monitor.
   The Job Monitor panel will open. If you completed Exercise 3.1 and enough time has passed (about 30 minutes), then you should see an entry with a description of "Bimsville Bypass profile optimization" with a green check mark in the Status column. If not, further instructions will follow.

8. Click the icon in the Report column. If you do not have access to this icon, you can open `Bimsville Bypass.pdf` located at `C:\InfraWorks Roadway Essentials\Chapter 03\Optimization Results\`.
   A PDF of the report will open. Notice that the new profile has 11 total PVIs. You started with 2.

9. Study the report and close it when you are done.

10. In the Job Monitor panel, click the icon in the Result column. If you have not completed Exercise 3.1 or if your results are not available, set Ex_3_1_and_3_2_End as the current proposal. Read through the next steps but don't perform any of the actions until you get to step 19.

You will see a dialog indicating that a file is downloading. Then a warning dialog will open indicating that the alignment has changed since the optimization was performed.

11. Click Yes to dismiss the warning dialog. Click Yes to create a new proposal.

12. In the Add New Proposal dialog, enter **Optimized_Bimsville_Bypass** in the Name field. Click OK.
    Optimized_Bimsville_Bypass will now be your current proposal.

13. Change the current proposal back to Ex_3_1_and_3_2.

14. Click the InfraWorks Home icon.

15. Click the InfraWorks Manage icon.

16. Click Proposals.
    The Proposals panel will open.

17. In the Proposal panel, click Merge Proposals.

18. In the Merge Proposals dialog, click Optimized_Bimsville_Bypass and then click OK.
    After a pause, the model will update and the Bimsville Bypass road will be replaced with the new design.

19. The Profile View panel should still be open. If it is not, open it by clicking Profile View on the Analyze toolbar.

20. Click the newly imported road to view its profile in the Profile View panel. This version of the profile has many more PVIs.

21. Study the profile and take note of how the optimization process redesigned it (see Figure 3.11).

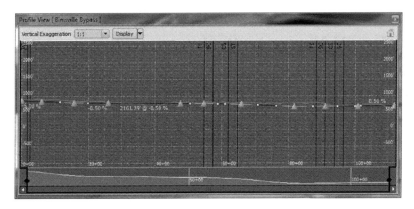

FIGURE 3.11 An optimized profile

Compare this profile with Figure 3.7. Notice how the slopes are all between 0.5 percent and 5 percent. These were the values you provided for Required Drainage Grade and Maximum Grade. Notice also that there is always at least 500 feet (150 meters) between PVIs because of the Minimum PVI Spacing value you provided.

Analyzing Sight Distance

Of course, one of the most important aspects of road design is driver safety. And one of the best ways to keep drivers safe is to give them good visibility. For that reason, a sight distance analysis is an important part of every road design. In the past, sight distance analyses have been time-consuming and expensive. But now you have an engineering-accurate 3D model and the power of InfraWorks. It has become almost *too* easy.

To perform a sight distance analysis, you will need the Sight Distance panel, which you open by clicking Sight Distance Analysis on the Analysis toolbar (see Figure 3.12). You can perform a sight distance analysis for two scenarios: a roadway and an intersection.

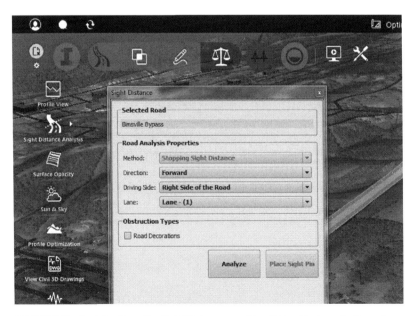

FIGURE 3.12 Open the Sight Distance panel by clicking the icons in the order shown.

Analyzing a Roadway for Sight Distance

If you choose a road for sight distance analysis, you will be presented with the following options on the Sight Distance panel (see Figure 3.13):

Road Analysis Properties   This section is used for setting up the configuration of the analysis.

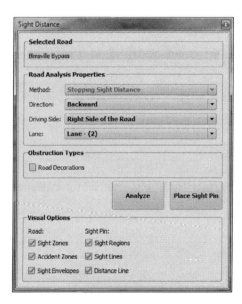

FIGURE 3.13   The Sight Distance panel for roads

Method   The method can be Stopping Sight Distance or Passing Sight Distance. For roads that have multiple lanes in either travel direction, only the Stopping Sight Distance method is available.

Direction   The direction is determined by the way you drew the road when you created it. If you drew it south to north, then the northbound lanes are Forward, and the southbound lanes are Backward.

Driving Side   Choose the driving side that matches the country you are in. For example, you would choose Right Side Of The Road for the United States and Left Side Of The Road for the United Kingdom.

Lane   For multilane roads, choose the lane you want to analyze. This works outward from the centerline, so for example, if there were two northbound lanes, the inside lane would be Lane - (1), and the outside lane would be Lane - (2).

Obstruction Types   If you want the sight distance analysis to consider 3D objects in the model, you can check the box next to Road Decorations.

Visual Options   The analysis shows results along the road using a number of colored zones and envelopes (see Figure 3.14). In this section you can choose which indicators are visible and which are not.

FIGURE 3.14   Sight distance analysis indicators for roads

Road: Sight Zones   Sight zones are colored lines that appear along the lane you've selected. They indicate Sight Clear areas (light greenish-blue) or Sight Failure areas (yellow).

Road: Accident Zones   Accident zones are areas that are especially prone to accidents because of poor sight distance. These are shown as dark greenish-blue lines within the lane you're analyzing.

Road: Sight Envelopes   Sight envelopes are light blue areas where obstructions are analyzed and, if they're found, are highlighted in orange.

You can also place a sight pin on your analysis: a single point from which you assess sight distance. There are visual indicators for sight pins too (see Figure 3.15).

FIGURE 3.15   Sight distance analysis indicators for sight pins

Sight Pin: Sight Regions   This is the region that is analyzed for obstructions. It is shown in several colors with orange representing obstructions.

Sight Pin: Sight Lines   If no obstruction is present, a sight line is a straight line from the sight pin to the end of the sight distance. If there is an obstruction, additional lines will appear indicating the first and last lines along which the obstruction is encountered.

Sight Pin: Distance Line   This is a black line with an arrowhead at the end that follows the path of the lane. The tip of the arrow indicates the required sight distance from the sight pin.

Exercise 3.3: Analyze a Road's Sight Distance

In this exercise you will perform a sight distance analysis for a section of the Bimsville Bypass.

   If you are continuing from the previous exercise, you can skip to step 3. Otherwise, if you haven't already done so, go to the book's web page at www .sybex.com/go/roadwayessentials and download the files for Chapter 3. Unzip the files to the correct location on your hard drive according to the instructions in the introduction.

1. If it is not already open, launch InfraWorks 360.

2. On the Start Page, click Open and browse to C:\InfraWorks Roadway Essentials\Chapter 03\. Click Ch03 Bimsville Roads.sqlite and click Open.

3. On the Utility Bar at the top right of your screen, click the Proposals drop-down list and select Ex_3_3_and_3_4.

4. Restore the bookmark named Road SDA.

5. Click the Roadway Design icon to expand the Roadway Design toolbar.

6. On the Roadway Design toolbar, click the Analysis icon to open the Analysis toolbar along the left side of the InfraWorks window.

7. On the Analysis toolbar, click Sight Distance Analysis.
   The Sight Distance panel will open.

8. Click the Bimsville Bypass road.
   The functions on the Sight Distance panel will activate. Notice that the only available choice for Method is Stopping Sight Distance. Passing Sight Distance is not available because this road has multiple lanes.

9. On the Sight Distance panel, do the following:

   ▶ For Direction, select Backward.

   ▶ For Driving Side, verify that Right Side Of The Road is selected.

   ▶ For Lane, select Lane - (2).

   ▶ Click Analyze.

   After a pause, indicators from the analysis should appear in the model.

10. Under Visual Options, check all of the options under Road.

   You should see various colored areas representing different key findings for the analysis.

11. Hover you cursor in the yellow area.

   Note the tooltips that indicate the Sight Failure and Accident Zone areas, as shown in Figure 3.16. The failure is caused by the embankment (highlighted in orange) that is obstructing a driver's view of the required sight distance.

FIGURE 3.16 Sight Failure and Accident Zone areas revealed by a sight distance analysis

12. In the Sight Distance panel, click Place Sight Pin. Click a point within the yellow Sight Failure area.

13. In the Sight Distance panel, within the Visual Options section, uncheck all boxes under Road and verify that all boxes under Sight Pin are checked.

14. Zoom in and study the graphics representing the sight pin analysis, as shown in Figure 3.17.

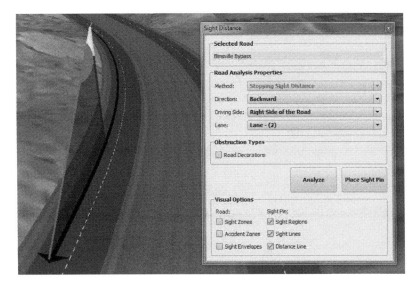

FIGURE 3.17   Analyzing sight distance using a sight pin

You should see two sight lines: one projecting to the end of the sight distance (the end of the black arrow) and the other passing through the first point obstructed by the embankment. The zone between them represents the sight region where the obstruction has an effect. You should also see the embankment highlighted in orange indicating the source of the obstruction.

Since there was no change to the model, there is no Ex_3_3_ and_3_4_End proposal for this exercise.

Analyzing an Intersection for Sight Distance

An intersection is another key location where sight distance is especially important. The Sight Distance panel handles intersections a bit differently, and as a result, different options are presented when you select an intersection (see Figure 3.18).

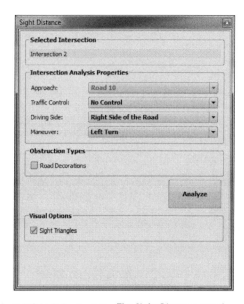

FIGURE 3.18   The Sight Distance panel
for an intersection

Intersection Analysis Properties   In this section you will configure the
analysis.

> **Approach**   For four-way intersections, you can choose which
> approach road you are analyzing and in which direction. For three-
> way intersections, this is always the "stem" of the T.
>
> **Traffic Control**   Here you can choose from No Control, Stop
> Control, and Yield Control. Each one has different sight distance
> requirements. No Control means that traffic will travel right
> through the intersection without regard for the other intersecting
> roads. Stop Control and Yield Control are self-explanatory.
>
> **Driving Side**   Choose the driving side that matches the country
> you are in. For example, you would choose Right Side Of The Road
> for the United States and Left Side Of The Road for the United
> Kingdom.
>
> **Maneuver**   Choose from Left Turn or Right Turn.

Obstruction Types   Use this section to choose whether you want road decorations (signs, hydrants, and so on) to be analyzed.

Visual Options   For now, this section has only the Sight Triangles option, which is the only visual indicator available. Perhaps there will be more options here in the future.

When you run the analysis, you will see the results as visual indicators (see Figure 3.19). Sight triangles indicate the required visibility as a vehicle approaches the intersection. If the triangle is light blue, the approach is clear; if it is yellow, the approach is obstructed. An arrow showing the path of the vehicle will be colored yellow if there are any sight distance issues and will be colored light blue if it is all clear. Obstructions will be highlighted in orange.

FIGURE 3.19   An intersection sight distance analysis

Exercise 3.4: Analyze an Intersection's Sight Distance

In this exercise you will perform a sight distance analysis for an intersection within the industrial park. You will correct a sight distance issue by performing some simple grading.

If you are continuing from the previous exercise, you can skip to step 3. Otherwise, if you haven't already done so, go to the book's web page at www .sybex.com/go/roadwayessentials and download the files for Chapter 3. Unzip the files to the correct location on your hard drive according to the instructions in the introduction.

1. If it is not already open, launch InfraWorks 360.

2. On the Start Page, click Open and browse to C:\InfraWorks Roadway Essentials\Chapter 03\. Click Ch03 Bimsville Roads.sqlite and click Open.

3. If it has not already been done in the previous exercise, on the Utility Bar at the top right of your screen, click the Proposals drop-down list and select Ex_3_3_and_3_4.

4. Restore the bookmark named Intersection SDA.

5. If the Sight Distance panel is already open, you can skip to step 8. Otherwise, click the Roadway Design icon to expand the Roadway Design toolbar.

6. On the Roadway Design toolbar, click the Analysis icon to open the Analysis toolbar along the left side of the InfraWorks window.

7. On the Analysis toolbar, click Sight Distance Analysis.
The Sight Distance panel will open.

8. Click the intersection shown in the current view.

9. In the Sight Distance panel, do the following:

   ▶ For Traffic Control, verify that No Control is selected.

   ▶ For Driving Side, verify that Right Side Of The Road is selected.

   ▶ For Maneuver, verify that Left Turn is selected.

   ▶ Click Analyze.

Notice that both sight triangles are yellow because of the obstructing embankments (see Figure 3.20).

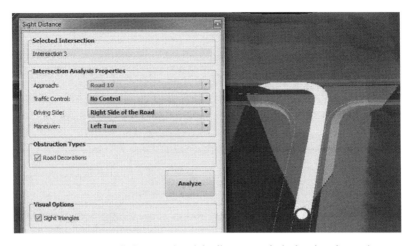

FIGURE 3.20 An intersection sight distance analysis showing obstructions on both sides

10. In the Sight Distance panel, change Traffic Control to Stop Control. Click Analyze.

    With this configuration, the analysis is all clear. The sight distance requirements are greatly reduced because the analysis considers the use of a stop sign in this scenario.

11. Click the embankment to the near right of the intersection. If the gizmos for the small coverage in this area do not appear, right-click and select Edit.

    You should see the gizmos of a small coverage area, as shown in Figure 3.21.

FIGURE 3.21 A coverage that will be used to grade the area adjacent to an intersection

12. Right-click the coverage and select Shape Terrain. Click the blue arrow gizmo and type 745 (227) in the tooltip next to Elevation. Press Enter.

This flattens out the embankment area ideally removing the obstruction.

13. Click the intersection. On the Sight Distance panel, for Traffic Control, select No Control. Click Analyze.

This time, with the embankment flattened because of the grading, the right sight triangle is all clear, as indicated by the light blue coloring (see Figure 3.22).

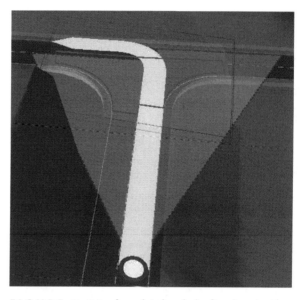

FIGURE 3.22   An updated analysis after changing the design

You can view the results of completing this exercise successfully by selecting the Ex_3_3_and_3_4_End Proposal.

Creating Civil 3D Drawings

As you have seen, Roadway Design for InfraWorks 360 is an amazingly powerful tool for creating, visualizing, and analyzing your road designs. Even so, the detailed design is still in the hands of AutoCAD Civil 3D. The Civil 3D environment is very different from InfraWorks. It is built on AutoCAD and uses the DWG format. To easily go from InfraWorks to Civil 3D, a translation tool is a must—and the View Civil 3D Drawings command answers this call.

You can find this tool on the Analyze toolbar; when launched, it opens the Create Civil 3D Drawings dialog. This dialog is in the form of a wizard and has three views: Select A Model Road, Select Surface, and Specify Civil 3D Options. The following sections outline the details of the Create Civil 3D Drawings dialog, broken down by view.

The Select A Model Road View

In this view you will configure the Plan Production aspects of the export (see Figure 3.23). What you choose here will determine details such as sheet size and scale that will make your final plans look the way you want them to look.

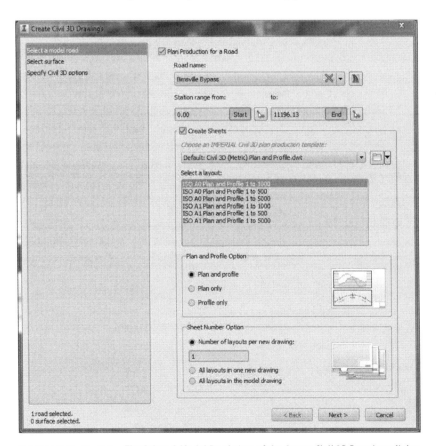

FIGURE 3.23 The Select A Model Road view of the Create Civil 3D Drawings dialog

Plan Production For A Road   Plan Production is a Civil 3D feature that automatically configures and creates the plan and profile sheets for a road. It is a complex feature that automates many Civil 3D and AutoCAD functions. If you're not familiar with this feature, you should definitely read up on it. With this option turned off, the function basically becomes an IMX export that also creates a DWG file.

> **Road Name**   Click the Pick A Design Road button and then click your road to select it. You may need to create a proposal with a simplified version of your road so that you can export it using this feature.
>
> **Station Range From/To**   You can restrict your export to a portion of your road, rather than the whole thing. You can supply the station range numerically or by using the pick buttons to choose the start and end locations.
>
> **Create Sheets**   In this section you will configure precise details for your sheets such as template, sheet size, layout (plan & profile, plan only, or profile only), and how to distribute the resulting layout tabs (all in one drawing file or distributed across multiple drawing files). These options match the choices in Civil 3D closely, so if you are familiar with Civil 3D Plan Production, this will be intuitive.

The Select Surface View

In this view you will configure the means by which your surface area will be selected (see Figure 3.24).

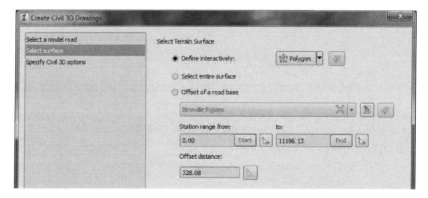

F I G U R E   3 . 2 4   The Select Surface view of the Create Civil 3D Drawings dialog

There are three options:

Define Interactively   This is a selection method you are likely already familiar with because it appears elsewhere in InfraWorks. With this option you can sketch an area by polygon, rectangle, or bounding box.

Select Entire Surface   Just like it sounds, this option will choose the entire surface represented by your model extent. This choice is not recommended for most models, especially large models. It could cause the export process to take a long time and could generate a large amount of surface data in areas that you don't need.

Offset Of A Road Base   With this option, you define the surface area by offsetting the centerline of your design road (see Figure 3.25). You can define a station range and offset distance. You can even click the Edit Area Of Interest button to make adjustments to the area after it has been defined by your parameters.

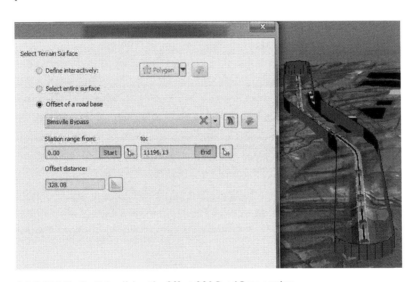

FIGURE 3.25   Using the Offset Of A Road Base setting

The Specify Civil 3D Options View

In this view you will make some choices relating to the template file and output files that are associated with the export process (see Figure 3.26).

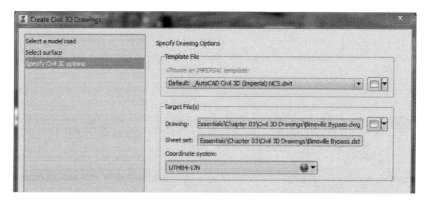

FIGURE 3.26   The Specify Civil 3D Options view of the Create Civil 3D Drawings dialog

Template File   Here you will choose the template that is used for your main output drawing file. The sheet files that are created will xref this file.

Target File(s)   The export process will create DWG files as well as a DST file for the sheet set that will be generated. In this section you will choose the location and filename for each one. Also, you can choose the coordinate system that will be applied to the drawing files.

The Final Result

Once you have chosen all of your options in the Create Civil 3D Drawings dialog, you can click Generate and launch the process. The processing time will vary depending on the length of the road and the detail of your terrain surface. Once the process is complete, you will see a dialog that will allow you to view the drawings in Civil 3D or open the folder where they have all been stored (see Figure 3.27).

FIGURE 3.27   The dialog indicating the completion of the process

If you click View Drawings, Civil 3D will open (provided you have Civil 3D installed on your computer), and you will begin at the master drawing. You will also see the Sheet Set Manager palette where you can double-click individual sheets to open them. If you open one of the individual sheets, you should see a fully prepared sheet complete with a layout, title block, viewports, xrefs, data references, and many other aspects that have been automatically configured for you (see Figure 3.28).

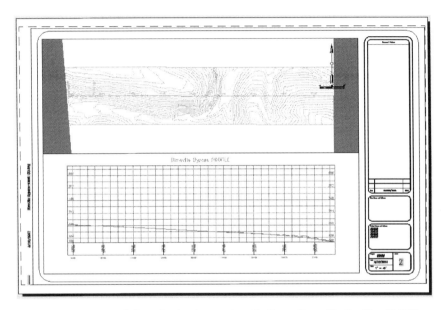

FIGURE 3.28   A sheet that has been automatically configured by the View Civil 3D Drawings command

From here you can continue to refine the design of your road, provide annotations, and eventually turn out a set of construction documents. The View Civil 3D Drawings command will give you a significant head start on the detailed design process.

Exercise 3.5: Generate Civil 3D Drawings

In this exercise you will generate plan and profile drawings for a section of the Bimsville Bypass road.

If you are continuing from the previous exercise, you can skip to step 3. Otherwise, if you haven't already done so, go to the book's web page at www .sybex.com/go/roadwayessentials and download the files for Chapter 3. Unzip the files to the correct location on your hard drive according to the instructions in the introduction.

1. If it is not already open, launch InfraWorks 360.

2. On the Start Page, click Open and browse to C:\InfraWorks Roadway Essentials\Chapter 03\. Click Ch03 Bimsville Roads.sqlite and click Open.

3. On the Utility Bar at the top right of your screen, click the Proposals drop-down list and select Ex_3_5.

4. Restore the bookmark named Bimsville Bypass.

5. Click the Roadway Design icon to expand the Roadway Design toolbar.

6. On the Roadway Design toolbar, click the Analysis icon to open the Analysis toolbar along the left side of the InfraWorks window.

7. On the Analysis toolbar, click View Civil 3D Drawings.
   The Create Civil 3D Drawings dialog will open.

8. In the Create Civil 3D Drawings dialog, click Select A Model Road and click Pick A Design Road next to Road Name.

9. Click the Bimsville Bypass design road in the model.

10. Under Create Sheets, choose Default: Civil 3D (Imperial) Plan And Profile.dwt if you are using imperial units. Choose Default: Civil 3D (Metric) Plan And Profile.dwt if you are using metric units.

11. Review the remaining settings and click Next.

12. In the Select Surface view, verify that Offset Of A Road Base is chosen under Select Terrain Surface.

13. Enter 300 (100) for Offset Distance and then click Next.

14. In the Specify Civil 3D Options view, under Template File, choose Default: _AutoCAD Civil 3D (Imperial) NCS.dwt if you are using imperial units. Choose Default: _AutoCAD Civil 3D (Metric) NCS.dwt if you are using metric units.

15. Under Target File(s), click the folder icon next to Drawing. Browse to C:\InfraWorks Roadway Essentials\Chapter 03\Civil 3D Drawings\ and click Save.

16. Click Generate.

The command will process for several minutes. When it is complete, a new dialog will appear giving you the options to view the drawings or open the folder where they are contained.

17. If you have Civil 3D installed on your computer, click View Drawings. Otherwise, click View Folder and note the files that have been created.

18. If you are able to open the drawings in Civil 3D, double-click each of the sheets on the Sheet Set Manager palette and study the output.

You should see sheets similar to Figure 3.29.

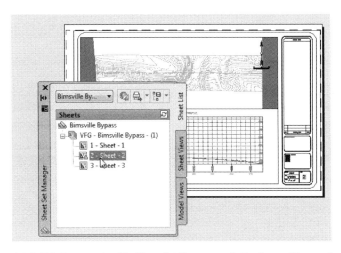

FIGURE 3.29 The Sheet Set Manager palette along with one of the sheets

You can view the results of successfully completing this exercise by viewing the files located at C:\InfraWorks Roadway Essentials\ Chapter 03\Civil 3D Drawings - Complete.

NOW YOU KNOW

Now that you have completed this chapter, you can make your InfraWorks roadway designs even better. You know how to optimize your profile to reduce costs while meeting your design requirements. You can also analyze roadways and intersections to identify sight distance issues or to show others that no issues exist. Finally, you are able to efficiently move your InfraWorks road designs to Civil 3D and get a head start on the detailed design process.

Now that you have completed this chapter, you can utilize the advanced roadway design optimization, analysis, and output tools available in Roadway Design for InfraWorks 360.

Index

Note to the Reader: Throughout this index **boldfaced** page numbers indicate primary discussions of a topic. *Italicized* page numbers indicate illustrations.

Made in the USA
Lexington, KY
06 April 2017